THE EPIDEMIOLOGY OF QUALITY

Vahé A. Kazandjian, PhD, MPH
Vice President for Research
Maryland Hospital Association, Inc.
Lutherville, Maryland
and
Assistant Professor
Epidemiology and Preventive Medicine
School of Medicine
University of Maryland
Baltimore, Maryland

with
Elizabeth L. Sternberg
Professional Writer
Rutland, Vermont

AN ASPEN PUBLICATION®
Aspen Publishers, Inc.
Gaithersburg, Maryland
1995

Library of Congress Cataloging-in-Publication Data

Kazandjian, Vahé A.
The epidemiology of quality/
Vahé A. Kazandjian with Elizabeth Sternberg.
    p.   cm.
Includes bibliographical references and index.
ISBN 0-8342-0533-5
1. Medical care—Quality control.
I. Sternberg, Elizabeth. II. Title.
[DNLM: 1. Quality of Health Care. W 84.1 K23e 1995]
RA399.A1K39   1995
362.1'068'5—dc20
DNLM/DLC
for Library of Congress
95-2642
CIP

Copyright © 1995 by Aspen Publishers, Inc.
All rights reserved.

Aspen Publishers, Inc., grants permission for photocopying for limited personal or internal use. This consent does not extend to other kinds of copying, such as copying for general distribution, for advertising or promotional purposes, for creating new collective works, or for resale. For information, address Aspen Publishers, Inc., Permissions Department, 200 Orchard Ridge Drive, Suite 200, Gaithersburg, Maryland 20878.

Editorial Resources: Jane Colilla

Library of Congress Catalog Card Number: 95-2642
ISBN: 0-8342-0533-5

Printed in the United States of America

1 2 3 4 5

I dedicate this book to my daughter Ani and my son Gregory, who fill my days with the comfort of being; and to my wife Janet, who made such fulfillment possible.

Table of Contents

Contributors	xi
Preface	xiii
Acknowledgments	xv

PART I—APPLICATIONS OF EPIDEMIOLOGICAL TECHNIQUES 1

Chapter 1—Quality: The Questions We Must Ask 3
Vahé A. Kazandjian and Elizabeth L. Sternberg

Incentives	4
Back to Consumers	6
Appropriateness and Efficiency	7
Value of Care	10

Chapter 2—Epidemiological Concepts in the Measurement of Health Care Quality 12
Haroutune K. Armenian and Vahé A. Kazandjian

Health Care Evaluation and Epidemiology	13
A Comparison of Rates	15
Methods of Investigation	21
Conclusion	23

Chapter 3—Indicators of Performance or the Search for the Best Pointer Dog 25
Vahé A. Kazandjian

The Most Popular and Misused Concept in Health Care	26

How Is an Indicator Defined after All?	27
The Making of an Indicator	28
Rate Construction	32
Errors of Construction and Interpretation	33
Conclusion	36

Chapter 4—Survey Methods for Quality Improvement Professionals ... 38
Michael A. Counte and Kristen H. Kjerulff

Survey Research Methods	39
Survey Research Applications	47
Conclusion	52

Chapter 5—Designing Quality Management Programs for Today and Tomorrow ... 55
Joann Genovich-Richards

Historical Background	55
Internal Organizational Assessment	61
Internal Organizational Evolution	75
Conclusion	81

Chapter 6—Comparative Performance Measurement for Health Plans ... 84
Janet M. Corrigan and Lisa S. Rogers

Characteristics of Health Plans	85
Menu of Performance Measures	86
Types of Performance Measures	88
Common Data Sources	93
Comparisons of Health Plan Performance	96
Public Reporting of Performance Data	100
Conclusion	103

Chapter 7—Closing the Loop: Are Clinical Pathways Our Most Comprehensive Approach To Measuring and Enhancing Performance? ... 107
Jonathan T. Lord

Strategic Approaches to Quality	108
Importance of Team Effort	109
Designing Care	110

| | Measuring and Assessing Performance .... | 119 |
| | Conclusion .... | 119 |

PART II—REPORTS FROM THE FIELD **121**

Chapter 8—Hospital Use of Clinical and Organizational Performance Indicators **123**
Anne M. Warwick, Anita M. Langford, and Judy A. Reitz

	Views of Quality	125
	Process vs. Outcome Indicators	126
	Performance Indicator Development	127
	Indicator Data Evaluation	132
	Communication of Indicator Information	140

Chapter 9—Collecting and Reporting of Patient Outcomes **144**
A.J. Harper

	Implementation	146
	Methods of Collecting and Reporting	148
	Responsibilities of the Report Format Task Force ...	158
	Verification Process	164
	Application of the Database	166
	Conclusion	168

Chapter 10—The Measurement and Management of the Quality of Ambulatory Services: A Population-Based Approach **171**
Norbert Goldfield

	The Importance of Continuous Quality Improvement	172
	The Importance of a Population-Based Community Perspective	173
	Changes in the Concepts of Space, Time, and People	174
	The Tools To Measure Ambulatory Care Quality ...	179
	Conclusion	191
	Appendix 10-A	194

Chapter 11—Approaches to Quality Improvement in the British National Health Service **197**
Liam J. Donaldson and Sir Donald Irvine

| | Historical Background of the National Health Service .... | 197 |

> The NHS of the 1990s: A System to Improve
> Quality 200
> Quality Improvement: Methods and Influences 201
> Assessing Service Performance 211
> Conclusion 220

Chapter 12—The Public Health Paradigm **222**
C. Patrick Chaulk

> Tuberculosis As a Public Health Problem 224
> Public Data and Public Policy on Tuberculosis 226
> The Tuberculosis Curve in the Age of AIDS 228
> Public Health Data and Program Evaluation:
> A Case Study 234
> Conclusion 235

Chapter 13—Measurable Accountability in an Era of Health Care Reform **239**
Kathleen R. Ciccone and Onita D. Munshi

> New York State Cardiac Surgery Reporting
> System 241
> Audience for the CSRS Data 244
> Lessons Learned 248
> Unresolved Issues 250
> Conclusion 251

Chapter 14—Quality and Community Accountability: A View of the American Hospital Association **253**
Thomas Granatir

> Community Care Networks: A Vision of Public
> Accountability 254
> A Framework for Community Accountability 256
> A Community Health Improvement Cycle 259
> Models of Community Health Improvement 262
> Major Challenges 264
> Conclusion 266

Chapter 15—From Theory to Practice: Managing Quality under Cost Constraints **269**
Timothy C. McKee and Timothy J. Ward

> Theoretical Model of Appropriateness 270
> Unified Model 273
> Conclusion 295

Chapter 16—Beyond the Hospital Door **297**
Vahé A. Kazandjian and Elizabeth L. Sternberg

    Common Methods of Inquiry 297
    State Level Initiatives 299
    National Initiatives 300
    Reorganizing through Health Care Networks 301
    Closing Thoughts 302

Index **305**

Contributors

Haroutune K. Armenian, MD, DrPH
Professor of Epidemiology
School of Hygiene and Public Health
The Johns Hopkins University
Baltimore, Maryland

C. Patrick Chaulk, MD, MPH
Assistant Professor
Department of Medicine
The Johns Hopkins Medical
  Institutions
Baltimore, Maryland

Kathleen R. Ciccone, RN, MBA
Vice President
Quality Assurance
Healthcare Association of New York
  State
Albany, New York

Janet M. Corrigan, PhD, MBA
Vice President
Planning and Development
National Committee for Quality
  Assurance
Washington, D.C.

Michael A. Counte, PhD
Professor and Chair
Department of Health Administration
Saint Louis University
Saint Louis, Missouri

**Liam J. Donaldson, MSc, MD,
  FRCS(Ed), FFPHM**
Regional General Manager
Director of Public Health
Northern and Yorkshire Regional
  Health Authority
Professor of Applied Epidemiology
University of Newcastle-upon-Tyne
Newcastle-upon-Tyne,
  Northumberland
England

Joann Genovich-Richards, RN, PhD
Assistant Vice President
Planning and Development
National Committee for Quality
  Assurance
Washington, D.C.

Norbert Goldfield, MD
Medical Director
3M/Health Information Systems
Wallingford, Connecticut

Thomas Granatir
Hospital Research and Educational
  Trust
Chicago, Illinois

A.J. Harper, MBA, RRA
Vice President
Professional Services
Greater Cleveland Hospital
  Association Professional Services
Cleveland, Ohio

Sir Donald Irvine, CBE, MD, FRCGP
Regional Advisory in General Practice
Postgraduate Institute for Medicine
  and Dentistry
University of Newcastle-upon-Tyne
Newcastle-upon-Tyne,
  Northumberland
England

Vahé A. Kazandjian, PhD, MPH
Vice President for Research
Maryland Hospital Association
Lutherville, Maryland
Assistant Professor
Epidemiology and Preventive
  Medicine
School of Medicine
University of Maryland
Baltimore, Maryland

Kristen H. Kjerulff, PhD
Assistant Professor of Epidemiology
Department of Epidemiology and
  Preventive Medicine
University of Maryland School of
  Medicine
Baltimore, Maryland

Anita M. Langford, RN, MS
Administrator
Johns Hopkins Geriatrics Center
Johns Hopkins Bayview Medical
  Center
Baltimore, Maryland

Jonathan T. Lord, MD
Executive Vice President
Anne Arundel General Healthcare
  System
Annapolis, Maryland

Timothy C. McKee, PhD
Principal
Health Services Engineering
Cabin John, Maryland

Onita D. Munshi, MBS, MS
Assistant Director
Quality Assurance
Healthcare Association of New York
  State
Albany, New York

Judy A. Reitz, MSN, ScD
Executive Vice President/Chief
  Operating Officer
Johns Hopkins Bayview Medical
  Center
Baltimore, Maryland

Lisa S. Rogers, MHS
Senior Associate
Barents Group LLC
KPMG Peat Marwick LLP
Washington, D.C.

Elizabeth L. Sternberg
Professional Writer
Rufland, Vermont

Timothy J. Ward, MS, PhD Candidate
Major
Biomedical Science Corps
Office of the Assistant Secretary of
  Defense
United States Air Force
Falls Church, Virginia

Anne M. Warwick, RN, MS
Director
Quality Services
Johns Hopkins Bayview Medical
  Center
Baltimore, Maryland

Preface

When we thought about a book on the quality of health care, we wanted a book that was different—not only in its focus, but also in the way the message was delivered. We wanted to synthesize, from the most prominent theories, applications, and proposed models of the twentieth century, what appeared to be the most innovative and the most effective approaches.

Today, the emphasis on improvement, change, cost consciousness, and social responsibility makes a transition to twenty-first century quality care less than smooth. Great things are happening, to be sure, but they are occurring amidst a tidal wave of change. We have asked experts to share with us their thoughts and experiences on the basics of performance assessment, evaluation, and monitoring. Thus, this book should be of interest to all in pursuit of quality of care.

WHY THE TITLE?

Because epidemiology is the study of what "falls upon populations," the epidemiology of quality could be described as the study of what happens to populations when exposed to different aspects and doses of quality. Such a conceptual framework promotes an interesting opportunity to explore the ways in which the seemingly elusive and nebulous essence of quality can be quantified and even evaluated. Specifically, it is our intent to (1) illustrate how the epidemiological methods of inquiry can be applied to studies of the quality of care; (2) describe the changes in population health status that occur over time following exposure to quality; and (3) demonstrate the common frame of thought underlying both the study of disease and that of health status following exposure to shades of quality.

The conceptual framework of this book is straightforward and, we hope, very logical. We start with theory, navigate through concepts, and graduate to the experiences of successful national initiatives not only in quantifying aspects of performance, but also in incorporating solid components of societal accountability. New models of performance assessment and management, where issues of cost and access raise unavoidable research questions and policy concerns, are discussed.

In our collaboration with nationally and internationally recognized experts in the quality of medical care, we knew that each would tackle the issue somewhat differently. We also knew, however, that there would be areas of overlap and that there might be significant differences in definitions and evaluations within those overlapping areas. We have not made judgments on these differences. Rather, we leave the reader to discover and enjoy the diversity of options, choices, successes, and lessons gleaned from those involved in these soul-searching challenges.

It is not our goal to advocate for a single initiative. Such decisions should be specific to the environment where care is provided and where recipient and provider expectations are more focused. Our goal, instead, is to present a book where concepts survive the tides of change; a book where the reader finds respite from circumstance, and comfort in the development of philosophy.

Finally, we have tried, as best we can, to present the material as clearly as possible. Our goal is to make this book accessible to a variety of health care managers, not just those in the "quality trenches." While attempting to clarify and simplify each chapter, we also strove to keep the voice of each chapter author intact, allowing them to tell their own stories in their own style.

May you find this book a reliable companion through your search for quality and understanding of performance.

VAHÉ A. KAZANDJIAN, PHD, MPH
BALTIMORE, MARYLAND

Acknowledgments

We thank Ms. Laura Pimentel of the Maryland Hospital Association for her dedication and patience in the management of communication with contributing authors and the publisher. Without Laura's spirit of organized tracking, this book would have not reached the publisher on time.

Part I

Applications of Epidemiological Techniques

» Chapter 1 «

Quality: The Questions We Must Ask

Vahé A. Kazandjian and Elizabeth L. Sternberg

PATIENT A goes into the hospital for a routine inpatient procedure. She receives kind and loving care from the primary physician and the support staff. Her hospital room is pleasant, and the food is good and served promptly. After discharge, her insurance carrier pays her entire bill. Her verdict? Top-of-the-line, quality health care.

Patient B goes into another hospital for a routine procedure. Her physician is aloof, and the nurses never answer when she calls. The food is tasteless and almost always served cold. The curtains in her room are ragged, and the window overlooks the hospital parking lot. When she is discharged, she is irritated to find that she must pay some of the bill out of her own pocket; insurance does not cover it all. Her judgment? Poor care, a bad experience, just one more example of how health care is going to the dogs.

So Patient A received quality care, and Patient B did not, right? While Patient A's perception was that her care was excellent, the following events took place behind the scenes: a nosocomial infection that was quickly controlled, minor medication errors, and overbilling not recognized by her insurance carrier. Patient B's care, meanwhile, included none of those occurrences; in fact, the behind-the-scenes work was excellent. Laboratory results were communicated quickly, medication requests were processed accurately, and billing paperwork was completed efficiently. Now who received quality care?

Welcome to the definition dilemma. What is quality health care? These two examples illustrate one of the problems in defining quality in the health care setting. That is, the patient's perception of quality may be markedly different from the health care provider's perception.

A closer look at the examples raises even more questions about ways to define quality care. Were the procedures necessary in the first place? Were appropriate medications and therapies ordered? Were the patients placed at risk for problems unrelated to their original diagnoses? Were the problems that occurred behind the scenes in Patient A's case aberrations, or were they average and expected? The questions can go on endlessly.

The definition of quality health care remains elusive, despite the reams of articles, books, papers, and speeches on the subject. It is easier to say what quality health care is *not* than what it is. It is *not* providing services that put patients at risk for little benefit. It is *not* recommending procedures and medications with high price tags and questionable results. It is *not* making mistakes when there are no second chances. In fact, books have been written, television shows produced, and sensational reports broadcast that tell in horrific detail what quality health care is *not*.

Clearly, there is still a long way to go to (1) define quality in health care and (2) explain to the public how it is measured and evaluated. Health care providers who wrestle with the definition of quality health care may sometimes look longingly at their industrial counterparts. In industry, quality's ultimate arbiter is the consumer. While this is true in health care as well, the identification of quality in the health care field has layers of complications. As a health care consumer, Patient A gave her care an unqualified "thumbs up." Yet, what health care provider would do the same? Defining quality, thus, becomes intertwined with communicating to the public appropriate ways to measure quality and combining the different facets of evaluation into a comprehensive statement about the goodness of care. As health care becomes a major sociopolitical issue, public understanding of quality evaluation has become of crucial importance. Quality and accountability go hand in hand.

Before health care workers feel that the accountability spotlight is singling them out unfairly, they should take a look at the day's headlines. The communication industry has exploded. Every business, every organization, and every industry is under scrutiny to determine how its actions affect others. Turn to television news and see how backyard barbecues deplete the ozone layer. Switch on the radio and listen to how second-hand smoke creates a health hazard for everyone around it. Pick up *The New York Times* and read how the disposal of motor oil pollutes the local watershed. Nobody is immune to accountability, least of all the health care profession.

INCENTIVES

Accountability is not a new issue to health care, but it can be a prickly one. Quality in industry is "free," the saying goes, because it increases sales; thus, the incentive to provide quality service or a quality product is inextricably linked to the bottom line. In health care, the incentives are different and may even be perverse. For example, certain kinds of payment systems may actually increase the utilization of a service, irrespective of the appropriateness of the service. So financial pressures are not always enough to ensure quality care.

Many health care providers may describe their incentive (for quality) as a desire "to do good." This reference to altruistic motives often arises in discussions of health care. Unfortunately, history has countless stories in which good motivations have resulted in disaster. In health care, the list includes experiences with drugs such as thalidomide and diethylstilbestrol, as well as questionable procedures. Merely wanting to do good, while laudable, is not enough.

The souring of good intentions is the very reason that accountability is so important. Health care is subjective enough without those involved in it offering a smoke screen of "good intentions" to cover even well meaning mistakes. When searching for quality, it is sometimes important to divorce "feelings" and "desires" from the definition. To minimize the subjectivity of quality assessment, both from recipients and from providers of care, it is necessary to look for objective, duplicable, and testable measures of quality.

Quality is quantified all the time in industry. Even in service industries, measurements are used to determine such things as how long it takes to serve a customer and how many hours of training make for good service. Attention to seemingly small details can often make or break businesses. Peters and Waterman noted that rigid quality rules are applied in other industries.[1]

- At McDonald's, Big Mac's must be discarded after 10 minutes if not sold; French fries, after 7.[2]
- Even ticket-takers at Walt Disney World must undergo four 8-hour days of instruction before being allowed to go "on stage." (Disney refers to working with the public as being "on stage.")[3]
- At IBM, some of the best salesmen are made assistants to officers, where they spend their time doing only one thing—answering customer complaints within 24 hours.[4]
- Proctor & Gamble insists that its employees' memos be no longer than *one page*. The reason? Longer ones are difficult to grasp, analyze, and verify.[5]

All these approaches involve quantifying some aspects of quality, or quality programs, and applying strict parameters. In the case of

McDonald's, products held longer than a certain period of time are deemed unacceptable, and a time frame for discarding them has been determined. In the case of Walt Disney, customer service is deemed so important to the company's success that a minimum number of hours of training is considered necessary for even the most mundane job. In the case of IBM, the quality of customer service is defined partly in terms of the length of time that a customer should expect to wait to get results from a complaint. Finally, in the case of Proctor & Gamble, a past president determined that quality sometimes hinges on the ability of those within the company to communicate effectively with each other; the way to do that, he believed, was to require employees to abide by strict rules for written material.

BACK TO CONSUMERS

Before a business establishes strict quality rules, it ascertains what the consumer wants. In industry, according to Donabedian, "there is . . . continuing preoccupation with eliciting consumer requirements and translating them into product characteristics. In one influential formulation, the consumer becomes, in fact, part of the production line."[6] This emphasis on consumer satisfaction is the engine that drives the industry quality machine and sets the framework for initiatives to quantify and evaluate quality, such as those mentioned earlier.

The parallels and differences between industry and health care are significant. Consumer satisfaction is important in both, but no health care organization is likely to replace its quality assurance team with the marketing department. If consumers are theoretically part of the production line in industry, they are production line worker, input material into the production process, and end product in the health care field.

Patient A, for example, can become part of the production line by filling out a hospital marketing survey. Her role, as part of the production line influencing the attitude of the staff, quality of the food, and price of the care, is significant. In addition, however, she is the input material upon which the attentions of the other workers on the line have focused. As such, she has been hampered by a lack of technical knowledge. She cannot say, "Your sterile techniques are poor; therefore, I'm at risk for a nosocomial infection." She cannot point out a defect in the medication administration system that is leading to minor errors. She cannot see into the billing department to straighten out the difficulties there. Her ability to influence these aspects of her care is limited. She does not even know to judge them, let alone how to judge them. Patient A has also been the outcome of the production process. Her health status during and at the end of her exposure to health care was the most obvious proof of the production goodness. Or was it?

Another key difference between the industrial and health care models is the cost-effectiveness of quality. As pointed out earlier, quality pays in industry. In health care, however, there may be no immediate hope of "increased sales" to offset the often high costs of quality improvement. Health care, therefore, must be much more concerned "with the balance of costs and benefits than the industrial mode."[7]

Complicating health care's quest for quality is its inclusion in the "service industry" arena. Even business people acknowledge that measuring quality in these industries is harder than in manufacturing. "Our employees create it [quality service], and then it disappears," says a vice president for corporate quality improvement at Chicago & North Western Transportation Co.[8] Measurement is difficult under circumstances when there is no real product, but only the manner in which a task is performed. A marketing survey of Patient A and those like her would have given high marks to the institution that cared for her. Conversely, Patient B's institution may have been chagrined to learn the results of a marketing survey of its patients.

What can business teach health care providers about defining quality? Somewhere along the line in all the examples given, people established boundaries within which they found the quality of their product acceptable. Perhaps McDonald's French fries taste about the same after 8 minutes as they do after 7 minutes, but someone, somewhere, using some kind of system, determined that 7 minutes is the most *appropriate* time to keep French fries on hand. Perhaps complaints at IBM can be answered in 48 hours rather than 24 with nearly the same results; again, someone somewhere, perhaps using marketing survey results or focus group materials, decided that 24 hours is the most *efficient* time frame within which to address complaints in a way that ensures customer satisfaction.

Just as industry uses appropriateness and efficiency of the service production to answer the question, Are we doing the right things for our customers, and are we doing them well?—so, too, must health care providers be prepared to use two gauges to judge whether they are producing a quality product. Most health care workers are aware, on some level, of these challenges and ironies.

APPROPRIATENESS AND EFFICIENCY

In health care as in all things, appropriateness can be defined as "doing the right thing." The appropriateness of health care refers to clinical judgment, although the administrative aspects of care can and will also influence clinical management. It is practically impossible to examine traditional quality assurance in health care without focusing on things that are done wrong instead of those that are done right. In fact, that is the credo of quality assurance: Find the wrong things. No one would suggest that this type of inquiry stop; as noted earlier, defining quality *is* often a process of striking out the wrong things in the hope that only the right things will remain.

Sometimes knowing what is right, or some facet of what is right, makes it possible to determine what went wrong and to ameliorate care. This approach is difficult to put into action, however. For example, many people are quick to denounce Caesarean section rates in the United States as "wrong" because they are so high. Yet few have identified the "right" rate. In the case of Caesarean section, as well as in most health care services, the evaluation of appropriateness is an evaluation of the *process* of the care, not necessarily only its immediate outcome.

Even though it is easier to define what is wrong, the search for what is right should not be abandoned. The question of appropriateness must be answered first, because it represents a very basic reality. It is possible to do the wrong things very well. The wrong things occur primarily when the right things are not clearly and commonly known.

Doing the right thing (the question of appropriateness) rests on the assumption that the necessary conceptual understanding, tools, and methods are available. Appropriateness deals primarily with the adequacy of a course of action to respond to a patient's biological, psychological, and perhaps even behavioral needs. Clearly, the adequacy of those actions depends on the state of the health care provider's knowledge.

In days gone by, treatment with leeches was deemed appropriate. Since those times, the science of medicine has advanced in its ability to gauge clinical appropriateness, but it is not yet and may never be an exact science. Medicine, as a science, is based on a set of probabilities. The same course of action does not always produce the same result. Moreover, the health care delivery system has its own set of probabilities. Combined, the two sets of probabilities make for a substantial margin of error at times. The old saying, "The surgery went well—but the patient died," illustrates this point. On the other hand, the opposite is also sometimes true. Care and methods of treatment may be inadequate, yet the outcome may be successful, as in the case of Patient A. In medicine, therefore, it is possible to do the wrong thing and still have something right occur in the end.

Doing the right things must fit certain societal criteria as well. These criteria are often hotly debated in today's headlines. Should life support be extended indefinitely to comatose patients? Should Caesarean section be performed unquestioned in search of a perfect baby? Should a 79-year-old nursing home resident receive a hip replacement? Should all insurance carriers pay for in vitro fertilization for infertile couples?

Society needs a clearer understanding of the methods of care available and the influences on the choice of those methods. For example, many factors, even administrative ones that *have an effect on* clinical practice, can influence the number of Caesarean sections done. In small, rural hospitals, anesthesia services may not be available on weekends or at nights. In hospitals such as these, many Caesarean sections may be performed on Fridays to avoid a weekend emergency situation. Management's decision to organize its financial and human resources in such a configuration has an impact on method of care and, therefore, on appropriateness of care.

When the number of primary care physicians is greater than the number of obstetricians in an area, the number of Caesarean sections may be lower. Because of their training, primary care physicians may not be as quick to perform the procedure. The training of the physician, in fact, may be a prime influence on such decisions. Who was the physician's mentor? Where did he or she go to medical school? Most young obstetricians in the United States are not trained in turning a breech presentation, for example, as thoroughly as their senior colleagues may be. When there is a complication involving a breech presentation, these physicians, *because of their training*, would be inclined to perform a Caesarean section.

There is yet another possible scenario in areas where only one or two obstetricians are available to a largely dispersed rural population. There may be high rates of Caesarean section and wide variations over time. Curiously, the fewer number of obstetricians may lead to a lesser controlled, rarely peer-reviewed practice style. Thus, the relationship between the profiles of utilization of services and the availability of providers may not always be unidirectional. In fact, the distribution may be bimodal; there may be significantly different practice profiles at the spectrum ends.

Such oddly related aspects have a profound effect on appropriateness from both clinical and societal perspectives. In Ireland, the Caesarean section rate is approximately 5 percent, although the rate of adverse and undesirable outcomes for both mother and baby seems similar, if not lower than in the United States, where the Caesarean section rate is approximately 23 percent of all deliveries. There may be two contributing factors: Irish physicians are trained in more conservative case management methods, and the Irish health care system allows midwives to be the primary provider responsible for labor.

Only by diligently searching for such factors that influence care is it possible to answer the question, Are we doing the right things? Furthermore, it has become clear that the greater the quantification—about outcomes, about methods of treatment, about anything affecting the nature of care—the closer the definition of appropriateness from a clinical perspective. The cold, nonjudgmental realm of statistics can help address aspects of the quality question.

VALUE OF CARE

A working definition for quality permits a look at *value*—a combination of quality and cost. How do patients know that they are getting value from the most social of all services, medicine? What are each one's basic requirements to be fulfilled, to "feel good" about the receipt of that service? The question asked about value of health care, or about caring, should not be different from the questions asked about the purchase of an airline ticket. People know where they want to go, and they want to get there with the maximum of comfort, security, and speed. They also want to feel good about the captain of the ship, as well as about the crew, the support staff. They want to be sure that there is more to eat than 14.2-g bags of peanuts and that the freshly brewed decaffeinated coffee is indeed freshly brewed. In short, they want quality service for a price that they can afford. Thus, the definition of value is a simple equation: the additive relationship between quality and cost. Value is a judgment that incorporates both the subjective and the objective; it is an evaluation conducted repeatedly in daily life.

Cost is the easier reality to quantify, hence evaluate. Usually, cost is immediately contrasted to the availability of resources, and a decision is made about whether the resources can support the expenditure. When an individual or a group realizes that the resources are finite and that, in some cases, they are shrinking, the process becomes easier to conduct and defend. But what about quality? How should quality be assessed, described, and evaluated? Should the goal of the journey be simply to get there or to get there with comfort and enough resources left over for others to enjoy the new destination?

The question of quality is intrinsically linked to the question of value. These are the challenges with which society is struggling. Some, like the Oregonians, have based decisions about the use of resources on the relationship of cost and quality. Others have initiated health care reform, where cost, quality, and access are the main variables in the equation.

Now, back to the patients. Who received the best care, the quality care, the care with the most value—Patient A or Patient B? The answer? Neither. If customer satisfaction does not cover all the facets of appropriateness of care, then it is necessary to judge Patient A's care according to other criteria. By even modest standards, all the errors that occurred during her stay make her care suboptimal. Patient B, meanwhile, might find all the right marks showing up on a clinical and managerial checklist, but she was not satisfied because of the way the care was delivered—care with no "caring." Patient B's perception of quality is just as important as the clinician's perception.

Perhaps the quality of care evaluation methods of the future will involve the filling of an interactive grid, where the extent of congruence between the provider and recipient of care assessments is analyzed. Until then, we should never stop asking about the quality of care.

NOTES

1. T.J. Peters and R.H. Waterman, Jr., *In Search of Excellence* (New York: Harper & Row, 1982).
2. Ibid., 256.
3. Ibid., 167.
4. Ibid., 159.
5. Ibid., 150–151.
6. A. Donabedian, Continuity and Change in the Quest for Quality, *Clinical Performance and Quality Health Care* 1, no. 1 (1993):9.
7. Ibid., 10.
8. L. Armstrong and W.C. Symonds, Beyond "May I Help You?" *Business Week*, 25 October 1991, 100.

» Chapter 2 «

Epidemiological Concepts in the Measurement of Health Care Quality

Haroutune K. Armenian and Vahé A. Kazandjian

BETWEEN 1974 and 1978, the National Screening Service in Norway invited all women and men between the ages of 35 and 49 in three counties to have a physical examination. The 48,500 individuals who responded were followed through 1990. Follow-up examinations revealed 281 hip fractures. After analyzing the data, the investigators concluded that tall people had a threefold increase in risk for hip fractures compared to persons of the same sex who were at the short end of the distribution for body height.[1] Is it possible to conclude from this study that shorter people have a healthy advantage in the later ages?

The occurrence of two cases of rapidly progressive fibrosing interstitial nephritis in a nephrology practice prompted investigators from Brussels to search for similar cases in other nephrology centers in Belgium. They identified seven other cases with similar pathology. All nine patients had followed the same slimming regimen in a specialized clinic. During the previous 15 years of operation, however, there had been no similar problems in the clinic's patients. Had something changed? Yes. In May 1990, the clinic had introduced a therapy that involved the use of Chinese herbs. Although a chemical analysis of the herbs did not show nephrotoxic contaminants, "the medicinal preparation of the capsules taken by patients had different alkaloid profiles from those expected in Chinese plants."[2] Thus, the authors argued against the uncontrolled therapy with herbal preparations.

These two incidents illustrate the challenge of making appropriate inferences. Every day, the media bombard the public with information. It is often tempting, in this era of sound bites, to precipitously decide that olive oil is better for one's health than are other oils, that high doses of vitamin E prevent cancer, or that babies delivered through Caesarean section have a higher Apgar score. What are the processes to evaluate this information? What are the available decision-making tools?

HEALTH CARE EVALUATION AND EPIDEMIOLOGY

Mrs. X is admitted to the hospital for care of uterine leiomyoma. At every step of her care, decisions are made about the optimal course of management, and surgical treatment (i.e., hysterectomy) is prescribed. In the absence of unusual problems, accepted protocols are used for her care. Choices of the best options are based on a number of factors, including nonmedical factors.

As part of an ongoing quality assurance process, Mrs. X's medical record is reviewed retrospectively, and a judgment is made about the quality of care provided in the hospital. Underlying such a judgment are comparisons with accepted standards. In addition, other factors, which may be situation-specific or case-specific, are weighed.

In one component of the ongoing quality evaluation, the chief of the medical staff has evaluated the care given to hysterectomy patients. The evaluation is based on a review of the records of all patients who have undergone a hysterectomy over the past 2 years. Data from the review are compared with those from another hospital. Within the context of this evaluation, questions arise as to the appropriateness of the choice of the comparison hospital and the validity of some of the indexes used in comparison and evaluation.

Thus, at every step of the health care process—at the individual patient level and at the group level—decisions about the best courses of action are based on answers to specific problems posed by an individual case or on information generated as part of routine data collection. Health information systems provide tools for the collection and transformation of data into a format that is useful for decision making. Data obtained in the process of caring for an individual patient not only serve in the management of that particular patient, but also are helpful in evaluating the quality of care within the institution as a whole (i.e., when used as part of a review of data from a group of patients).

The judgments made about Mrs. X's care are about care in general at her facility. How the information that serves as the basis for the judgments is gathered is as important as the judgments themselves.

It is appropriate to look at epidemiology when searching for quality health care because epidemiology is an information science. It uses data generated through a variety of systems for decision making and evaluation. It provides the tools for making comparisons between various options; it has a set of processes for judgment that are very useful in making the appropriate inferences. Epidemiological investigators may use routine databases (e.g., the annual statistical reports from a hospital) to study trends over time, or they may initiate a special detailed study to resolve a perceived problem. Thus, in the absence of an established information system that provides data on utilization patterns, it may be possible to conduct a number of ad hoc studies of patients and medical records to answer questions for planning and management decisions. Even in the early developmental phases of a primary health care system, reviews of a sample of patient care records can reveal various profiles of utilization and other data that will help set priorities and influence resource allocation at the micromanagement level.

The word *epidemiology* itself highlights the important role of this discipline. The origin of the word (*epi*, 'upon' and *demos*, 'people') indicates epidemiology's role; it is a science that deals with the problems that befall people. The more traditional interpretation of epidemiology as the study of epidemics narrows its practitioners' role to one of dealing with *unusual* problems. Epidemics, after all, involve a level of disease or a health problem in the community that is greater than expected and, therefore, unusual. While it is true that epidemiologists deal with diseases as varied as cancer and salmonellosis, they also monitor human populations for health problems that are occurring more frequently than expected. Thus, problems of poor health care and poor outcomes are definitely issues that can benefit from epidemiological investigations. An epidemic in this context can be defined as a cluster of unexpected outcomes that require attention.

Epidemics are defined by three dimensions: place, time, and persons. For example, five infant deaths during a year in a developing country's hospital may not be considered an epidemic if the number of infant deaths has been similar over several previous years. The same number of infant deaths in a modern hospital in New York may be labeled an epidemic, however, because of the rarity of the event in that facility. In these cases, place is an integral part of the definition of an epidemic.

The identification of epidemics and the search for their causes are totally dependent on good information. Today, because of an increasing pressure to make health care decisions less subjective, there is a heightened emphasis on using well established epidemiological methods of health care evaluation. The underlying purpose is to evaluate the efficiency and effectiveness of programs. The development of health information systems that are now accessible to many health services professionals and organizations facilitates such evaluation, but the accessibility is a double-edged sword. Although technological advances in information systems have groomed a new breed of users, are these professionals skilled in the use of comparative methods and the organized inferential approaches of epidemiology? If not, they are apt to commit errors that may have serious impacts on human morbidity and health care cost.

A COMPARISON OF RATES

The basic instrument of comparison in epidemiology is the rate. It is a measure of the occurrence of an event. Written as a fraction, its numerator is a count of events; its denominator, the number of people at risk for that event. A mortality rate has a count of deaths in the numerator and the population within which these deaths were identified in the denominator.

> A rate in itself is not very informative unless it is compared to another rate.

The construction of a rate is as important as the appropriate application of the rate. Among the issues of construction are the rate's *validity* and *reliability*. Most issues of the validity of epidemiological data involve the accuracy of the numerator and the denominator. A review of the numerator of a nosocomial infection rate, for example, may focus on some of the following questions:

- How is a hospital-acquired infection defined?
- Are all such cases reported to the system?
- Are there some infections that are misclassified as hospital-acquired?

For the denominator, a review may concern such issues as whether the population at risk of infection includes patients hospitalized fewer than 48 hours or whether the count of hospital inpatients is appropriate. Other questions that may arise in a rate assessment include

- Are the persons in the numerator selected from the population in the denominator?
- Does the numerator include all the outcomes occurring in the denominator population?
- What are the resources and sources of data for identifying outcomes that are not accounted for in the calculation of these rates?

- Is there any variation in the way that the persons with the outcome have been identified and diagnosed? Specifically,
  1. Is there any misclassification in the numerator?
  2. Does the numerator include anybody who does not definitely belong there?
  3. What is the amount of such misclassification?
- In comparisons of multiple rates (among institutions, areas, or across time within the same hospital), are all rates calculated through uniform, standard methods? Are the outcomes identified and classified by means of a uniform method?
- Are there errors in the enumeration of the population at risk for the outcome under study in the denominator?

Incidence, Prevalence, and Measurement of Risk

Epidemiology uses a number of rates. The most useful for establishing relationships between factors and events are incidence and prevalence rates.

> An incidence rate estimates the dynamics of new occurrences of an event. It is usually expressed by using the number of new cases as the numerator and the population at risk as the denominator.
>
> A prevalence rate indicates the load of individuals with a particular disease in a community. It measures the number of all new and old cases of the disease during a particular time and is expressed by using the number of cases as the numerator and the population at risk as the denominator.

An incidence rate for 1993 of 243 appendectomies per 100,000 people in a community reflects the frequency of new operations in that year. The incidence rate is a measure of absolute risk. A member of this community had a 243/100,000 risk of an appendectomy during 1993. A prevalence rate of diabetes of 4 percent on July 1, 1993 in that same community indicates the load of diabetics that exist at a particular time.

The relationship between incidence and prevalence rates is very dynamic and is determined by the duration of the outcome or event in question. Thus, a hospitalization rate in a community is an incidence rate of persons admitted to the hospital during a unit of time. A census of beds occupied at one point in time provides the numerator for a prevalence rate of hospitalization in that same community. The length of hospitalization determines how high the prevalence rate is going to be. Patients with prolonged hospitalization have a higher probability of being captured as part of the numerator at more points in time. To visualize this dynamic relationship, one can imagine a system in which the inputs are the events of hospitalization or incidence, the outflow are discharges and deaths, and that which the system continues to hold is the prevalence or the load.

The use of these rates makes it possible to compare absolute and relative differences among various groups. A community may have 137 more appendectomies per 100,000 people per year than the national rate, for example. This is an absolute increase or the *attributable risk*. If a community has an appendectomy rate that is twice the national rate, this is a relative increase expressed as a *relative risk* (i.e., a ratio of two incidence rates). Thus, a statement that the relative risk of having an appendectomy in Community X is 2 reflects a ratio of the incidence rate of Community X and a comparison incidence rate. It indicates that the incidence in Community X is twice as high as that in the comparison group and that persons in Community X were at a twofold increased risk of having an appendectomy in 1993.

Validity and Reliability

Before any conclusions can be drawn from a comparison of rates, it is necessary to assess the validity and reliability of the rates. It is not unusual to find a large number of measurement problems as a result of either systematic or random errors. Systematic errors or biases are predictable, preventable, or correctable. Due to mistakes in the processes and measurement, they can occur at any phase of a study or at any stage of the process of data generation. For example, the use of differing definitions of nosocomial infection in two hospitals makes the rates from these two hospitals noncomparable. Not only are epidemiologists concerned with the identification of these biases, but also they have developed a number of approaches to prevent them or at least to assess the potential impact of these biases on the inferences that can be made.

Ascertaining the validity of a test (or indicator) involves comparing the measurement to a reference situation that is accepted as the standard. Validity is measured by *sensitivity* and *specificity*.

Sensitivity is a measure of a test's ability to identify outcomes correctly, while specificity measures its ability to recognize the absence of the outcome correctly. Exhibit 2–1 illustrates the relationship of sensitivity to specificity. Thus, the sensitivity of mammography is expressed as a percentage of the known breast cancer cases that the test is able to detect; its

Exhibit 2-1 Assessing the Validity of a Test

		Disease (Reference)	
		Present	Absent
Test	Positive	A	B
	Negative	C	D
	Total	A + C	B + D

$$\text{Sensitivity} = \frac{A}{A+C} \quad \text{Specificity} = \frac{D}{B+D}$$

		Cancer (Pathological Diagnosis)	
		Present	Absent
Screening Test	Positive	80	90
	Negative	20	810
	Total	100	900

$$\text{Sensitivity} = \frac{80}{80+20} = 80\%$$

$$\text{Specificity} = \frac{810}{90+810} = 90\%$$

specificity is the percentage of the patients with no cancer that the test is able to identify. Both of these measurements are calculated according to a standard of comparison; in this case, the histopathology serves as a standard of reference.

When there is no accepted standard of reference or when the relative merits of two processes are under investigation, measures of reliability may become as important as measures of validity. Reliability is determined by the number of random errors made in the process of information gathering or measurement. Given the same test conditions, how dependable is the process in providing similar results every time? Thus, a reliable measure can accurately compare the same process or person with itself or two processes or persons within similar situations.

Sources of Data

Ongoing collection systems, as well as special investigations, generate epidemiological data. With concerns about cost and efficiency of the health care system driving a number of decisions, most facilities have a plethora of data about the utilization of their services. These statistics make it possible to compare trends or to link the frequency of procedures and processes to a number of health outcomes.

A review of the annual statistics from the major hospital of the State of Bahrain revealed a seasonal variation in the number of deaths reported each month from the hospital.[3] Calculations of monthly fatality rates from monthly admission and discharge figures from the same report confirmed the initial observation that the mortality rate was higher during the winter months. Similar observations were made in other hospitals of the Middle East. An investigation initiated to determine the causes of the higher fatality rates linked a number of statistics to case fatality rates, including characteristics of the professionals, relative frequency of certain diagnoses and procedures, and the demographic characteristics of the patient population. In addition to these variables, investigators examined changes in the data-handling procedures and employee patterns. Although it would have been easier to point the finger of suspicion at any one of these factors, none fully explained the higher fatality rate in the winter months—except for the obvious climatological changes. Thus, it is important to consider a number of possible explanations for a variation in rates before reaching any conclusions.

Problem Definition

Defining a problem accurately is as important as searching for its solution. The first step in investigating the problem of nosocomial infections is to decide how to define it. Problem definition itself is based on a set of decisions about sensitivity and specificity. A sensitive definition may include all cases, but may not be very specific because infections due to other causes may be classified as nosocomial. The options available for interventions also influence decisions on sensitivity and specificity. Situations that are conducive to corrective action without costly intervention are the ones to emphasize sensitivity. For example, inexpensive blood tests for sexually transmitted diseases may well be worth the 5 percent error rate. A sensitive definition or test is the one that identifies the largest number of people with the problem.

System Characteristics

In addition to the issues of data validity and reliability, it is essential to address characteristics of the information system at every level of data collection.

Cost. There is an optimal level of efficiency in an information system; above a certain level of intensity of effort, efficiency is lost. It is very important to justify the cost of generating the information. Justification may be based on improved efficiency of the health care system in handling a larger load or eliminating waste. In addition to the monetary cost of generating the data, it is important to relate the value of the information to its impact on the process of decision making, as well as its possible effect on health indicators.

Timeliness. Unless the information precedes the decision, it is of no value to the users. A number of statistical systems fail to meet this criterion of timeliness. The frequency with which the data are provided is also important. Although timely, concurrent information is most likely to result in immediate action, retrospective data can provide valuable insight into the performance profile of an organization. Historical data can indeed identify patterns of outcomes or of processes that trigger corrective action, if necessary.

Content. The information must be appropriate in the context of decision making. No matter how good, the value of information is a function of its relevance to any issue being studied.

Quantity. A number of organizations limit access to useful data within the system for purposes of control. Also, a number of organizations spend a lot of resources to generate for the system massive amounts of information that either is unusable or confuses the professionals who are supposed to be the users. The goal is to generate the minimum necessary amount of information for the optimal decision. Providing hospital boards and medical staff with vast amounts of tabulated data, text, and crowded graphics does not impress users, it renders the exercise useless and can even foster a reluctance to request needed information in the future.

Coverage. Is the information reaching the potential users of the system and the sources providing data? A well prepared presentation of information is not enough; it must be presented to the appropriate audience. Traditionally, hospital trustees, management, and clinical professionals share the responsibility for health care. On a larger scale, the consumers of health care—both actual and potential—are also the recipients of information about a health care organization's performance. In the coming years, consumer education is expected to increase in prominence. The request for "data cards," comparative quality indicators, and profiles of patient satisfaction will become part of the routine data collection and information generation.

METHODS OF INVESTIGATION

In addition to routine data collection systems, a number of special studies are available for use in the decision-making process. Demographers, epidemiologists, and social scientists have developed such alternative approaches as special surveys, longitudinal studies, and case-control studies. Information obtained through such special studies can complement routinely collected data and can validate data from other sources. Longitudinal investigations start with a study population in which the outcomes (e.g., disease, disability) have not yet occurred and relate the frequency of the outcomes at some future date to various initial characteristics of persons in the population. The most common method of epidemiological investigation nowadays is the case-control method in which investigators begin studying persons both with and without the outcome and compare the two groups as to the frequency of antecedent characteristics. Both these approaches focus on identifying an association between a characteristic and an outcome, and establishing the antecedence of the characteristic to the outcome.

Longitudinal Methods

As stated earlier, the basic epidemiological approach aims at generating two rates that can be compared to each other. In a most direct approach, the rate of a particular outcome in those with a certain characteristic is compared with the rate of the same outcome in those with another characteristic.

> The longitudinal model is a direct way of testing the hypothesis of the relationship of cause and effect.

The experimental method is a longitudinal design in that the study population is allocated to two or more interventions and the incidence of the outcome is compared between the intervention groups at some future point in time. The rate of efficacy for a particular intervention is calculated by subtracting the incidence of the outcome in the group with the intervention from the incidence in the group without the intervention. It is thus possible to estimate the relative improvement in the presence of the intervention.

> The efficacy of aspirin in preventing myocardial infarction is measured by subtracting the incidence of these infarctions in those taking aspirin from the corresponding incidence rate in those not taking aspirin. Dividing this difference by the incidence of the disease in those not taking aspirin provides a measure of efficacy that indicates the level of relative protection from myocardial infarction in those taking aspirin.

In many instances, cost and ethical considerations make it difficult to embark on an experimental study. For example, if a study found that the transplantation of bone marrow benefited patients with a certain type of cancer, but cost $1 million per patient, should all those afflicted with that type of cancer receive that experimental treatment? Perhaps of more widespread relevance are the experimental protocols for infertility treatment, in vitro fertilization, and abdominal laser microsurgery for endometriosis. What are society's priorities in the allocation of resources and technology? An experimental study that exposes a group of persons to an untested biological agent or a chemical compound raises obvious ethical questions. Recently, such ethical concerns have been raised not only for human subjects, but also for laboratory animals.

In an observational study, the investigator does not intervene in the distribution of the intervention or characteristic of interest, but simply collects data on what is happening naturally. The investigator follows or traces a group of persons with a particular characteristic (i.e., a cohort) over time for the incidence of the outcome of interest and compares this incidence to the rate in another group that has a different characteristic or exposure. In this type of study, the investigator is able to calculate a relative risk, which is a ratio of two incidence rates.

> To study the role of different types of contact lenses in causing ulcerative keratitis, Poggio and associates surveyed ophthalmologists to identify keratitis cases in a five-state area in New England.[4] To provide denominators, they estimated the number of persons wearing different types of contact lenses in this same region from a household survey. "The annualized incidence of ulcerative keratitis was estimated to be 20.9 per 10,000 persons using extended-wear soft contact lenses for cosmetic purposes and 4.1 per 10,000 persons using daily-wear soft contact lenses."[5]

The relative risk for ulcerative keratitis of users of extended-wear soft contact lenses in this study was $20.9 \div 4.1 = 5.1$.

Case-Control Methods

It is not easy to embark on a longitudinal follow-up of a population in which effective numbers for the occurrence of the outcome of interest may be unobtainable for a number of years. Thus, a more efficient design is the case-control method, in which members of the study population are defined by the presence or absence of the outcome of interest. Evaluation of the frequency of the characteristics in both groups provides another measure of association: the odds ratio. This is the ratio of the odds of exposure in the case group (those having the outcome) to the odds of exposure in the control group (those not having the outcome).

> Schein and associates also conducted a case-control study to identify patterns of use that increased the contact lens user's risk for keratitis.[6] Eighty-six contact lens users with ulcerative keratitis were identified from six hospitals, and these were compared to two other control groups of contact lens wearers with no evidence of keratitis; 61 other patients were seen for unrelated reasons in the same hospitals, and a group of 410 persons from the community identified by a telephone survey. This study showed that the risk for ulcerative keratitis was increased an estimated fourfold in users of extended-wear lenses compared to daily-wear lens users. This study further distinguished the users by type of use. The use of the extended-wear lenses overnight increased the risk of keratitis 10 to 15 times in the various subgroups of comparison.

Over the past few decades, the case-control method has been used very extensively in a broad number of applications. From an initial use in etiological research, its uses have expanded to include program evaluation, ascertainment of efficacy of interventions, and screening.

CONCLUSION

Although there are a variety of sources of information and a number of methodologies to generate the necessary database for an evaluation of health care quality, the epidemiological approach has a well established theoretical and mathematical base with a wide scope of applications and should be useful in a number of situations. It is important to emphasize that the process of evaluation of quality must move away from its dependence on routine procedures that are centrally dictated by a number of problem-specific or situation-specific methodologies. A solid base in epidemiology enables the professional concerned with quality to ascertain the validity of the data, as well as to design the special investigations necessary to deal with specific problems. Programs of quality assurance must go beyond the routine of established systems; epidemiological methods will help achieve that goal.

Even with a routine information system, there is an element of discovery. Not every user will look at a database from the same perspective and same viewpoint. The discovery of new trends, differences, and new associations in a routine information report is a challenge for every health care professional working in quality. Epidemiological methods provide these professionals with a trained eye that will help make those discoveries.

NOTES

1. H.E. Meyer et al., Risk Factors for Hip Fracture in Middle-Aged Norwegian Women and Men, *American Journal of Epidemiology* 737 (1993):1203–1211.
2. J.L. Valherweghem et al., Rapidly Progressive Interstitial Renal Fibrosis in Young Women: Association with Slimming Regimen Including Chinese Herbs, *Lancet* 341 (1993):387–391.
3. H.K. Armenian et al., Seasonal Variation of Hospital Deaths in Some Middle-Eastern Countries, *Lebanese Science Bulletin* 4 (1988):55–64.
4. E.C. Poggio et al., The Incidence of Ulcerative Keratitis among Users of Daily-Wear and Extended-Wear Soft Contact Lenses, *New England Journal of Medicine* 321 (1989):779–783.
5. Ibid.
6. O.D. Schein et al., The Relative Risk of Ulcerative Keratitis among Users of Daily-Wear and Extended-Wear Soft Contact Lenses, *New England Journal of Medicine* 321 (1989):773–779.

» Chapter 3 «

Indicators of Performance or the Search for the Best Pointer Dog

Vahé A. Kazandjian

"ROOM SERVICE."
I opened the door; it was indeed room service (or rather it was a young woman ready to clean my room).

"I'm on my way out," I said. "If you could give me just another minute...."

I was at a hotel on a business trip. As I got ready to leave, I let in the young lady. I headed for the door but then the telephone rang. "Get started, if you like," I told her as I answered the telephone.

While I was on the telephone, she cleaned the bathroom, taking the towels out, putting in new little vials of shampoo and conditioner, and checking the shower cap that I never use. My call lasted a few minutes, but before leaving, I wished her a good day. While doing this, I noticed that she was replacing the tissue box in the bathroom, although I knew there were still tissues in the box. Curious, I asked her why she was replacing the box before it was empty.

"See," she said, showing me the box of white tissues. "You have reached the yellow sheets. That means that you will be running out of tissues within 20 or 25 sheets. It is time to replace the box."

I smiled and headed for the elevator. I was on my way to discuss clinical indicators at a conference. I now had a wonderful introductory anecdote to define an indicator.

THE MOST POPULAR AND MISUSED CONCEPT IN HEALTH CARE

Although indicators themselves are not a new concept, the potential usefulness of indicators of hospital performance has recently received wide attention because of the relatively new opportunities to create comparative databases regionally, nationally, and internationally. The definition of an indicator is simple: it is an observation expected to indicate a certain aspect of performance. In the case of the tissue box, the appearance of the yellow tissues prompted performance; it was a signal to the cleaning lady to change the box. The Dow Jones industrial average is an indicator that shows how many and what kinds of financial transactions took place during a business day. Furthermore, that average has a built-in value. Because a higher value indicates more extensive trading, it has a certain predictable influence on the performance of investments. In the field of sports, baseball provides the most comprehensive set of indicators regarding a team's or an individual player's performances. The number of runs batted in (RBIs), wins, losses, and home runs are all indicators on which to base decisions about and the "goodness" of the performance. Also, a knowledge of historical patterns and performance profiles permits intelligent guesses about future expectations. Thus, indicators can help establish two important parameters of decision making. First, attaching a value to the magnitude of the indicator makes it possible to determine the "goodness" of performance; second, the previous performance profiles of a well established indicator can help make predictions about future performance.

The most common use of indicators in health care is within the context of quality. Thus, it is necessary to determine the meaning of a "quality indicator." First, an indicator is expected to indicate, to point to, to reveal something. It seems logical to assume that a quality indicator will point to or indicate quality. Further, it seems likely that a particular indicator will be indicating good or bad quality. Of course, at this point, the indicators used in health care run into the perennial dilemmas concerning the very nature of quality. Quality in health care is not only in the eye of the beholder, but also in the purse of the beholder, in the expectations and hopes of the beholder, and in the interpretation of the beholder. Therefore, does society have impossible expectations of the usefulness of indicators in health care?

Indicators can be extremely useful tools in the search for quality, but only if they are applied correctly and their limitations are understood. When the tool yields results different from those expected, it may be more important to fine-tune the skills and understanding of the users of those tools than to blame the construction of the tools. Unfortunately, most efforts seem to be directed toward "perfecting" indicators instead of providing the necessary training for the users of those indicators and the interpreters of their results.

HOW IS AN INDICATOR DEFINED AFTER ALL?

There is only one surefire way to test an indicator—in the field. In fact,

> An indicator of performance is similar to a pointer dog taken to the field in search of a pheasant.

Indeed, the analogy of the pointer dog encompasses all aspects of the performance of an indicator.

The clear goal of hunting for pheasant is to find the pheasant and bring it home for dinner. This expedition involves two principal parties. The first is the pointer dog, a devoted companion whose job is to point in a certain direction where the pheasant is likely to be found. That pointer dog has been trained and chosen for effectiveness in detecting a "true" pheasant. The second party is the hunter, who is expected to achieve the goal of the expedition not only through an understanding of the dog's signals, but also through marksmanship capabilities. After all, a pointer dog is useless to a hunter who cannot aim accurately.

It is impossible simply to look into the eyes of a dog and determine whether the dog is going to be a good pointer dog. In fact, the only way to make such a determination is to take the pointer dog to the field. Even that is not enough, however. It is generally necessary to conduct a number of trials in different types of fields and conditions (e.g., where the brush is high or when the skies are cloudy). A good dog should work well in all terrains and atmospheric conditions.

When is a dog considered a good pointer? Is a dog that points correctly 50 percent of the time a good expedition partner? What about the remaining 50 percent, the false-positive incidents? Similarly, should a hunter keep a pointer dog with a 20 percent false-negative rate? Perhaps one hunter would decide that only a dog that is effective 100 percent of the time (pointing toward the bushes where a pheasant can be found) is a good pointer dog. Who makes the decision? What are the criteria for effectiveness? And what about the efficiency of the dog's performance? Perhaps the dog is effective a high percentage of the time, but covers the terrain in a manner that will exhaust the hunter. Therefore, the dog becomes a much less desirable companion. Finally, what if the pointer dog were taken to the field to hunt pheasant and pointed only to quails? How useful would the pointer dog be?

As stated previously, no matter how good the pointer dog, there would be no pheasant on the dining table if the hunter cannot aim and shoot accurately. At that point, whose fault is the lack of pheasant—the hunter's or the pointer dog's?

Like the "goodness" of the pointer dog, the "goodness" of an indicator is determined by taking it to the field and testing it over a period of time.[1] No group of experts or meta-analyses alone can decide the desirability of an indicator. Furthermore, decision making about the effectiveness and efficiency of an indicator resembles decision making about the false-positive or false-negative pointing percentage of the dog. Is an indicator "good" if 80 percent of the time it correctly reflects aspects of an institution's performance that could be modified and improved? What about 60 percent? Should the goal be perfection? Should only 100 percent effective indicators be used?

THE MAKING OF AN INDICATOR

There are two important questions to be answered in the making of an indicator. First, is the indicator truly capable of capturing or describing the feature of interest? In other words, is the indicator pointing to the right thing? Once this first question is positively answered, the second question is addressed. How well does the indicator point to the right thing? These questions translate directly into two fundamental statistical and conceptual aspects of any indicator development: validity and reliability.

There is considerable confusion in the field regarding the difference between the validity and the reliability of an indicator. The two terms are not interchangeable; they describe two complementary, but independent, characteristics of any measure.

The Validity Question

Much more conceptually focused than is reliability, validity depends on evidence that the indicator is measuring what it was intended to measure. For example, is the Caesarean section rate a valid indicator for the desirability and appropriateness of the procedure (Table 3–1)? Or is it an indicator of the availability of anesthesiologists on weekends? Or does that rate reflect the frequency of requests of highly educated women in their 30s who, expecting a perfect child from perhaps their only pregnancy, ask for the perfect baby without the pains of prolonged labor? These are difficult questions that transcend the very nature of indicators and encroach upon the question of interpretation of observed patterns and profiles.

Table 3–1 Caesarean Section Rate As an Indicator of Quality Care: Valid?

	Yes	Questionable
Obstetrician practice style (physician decision making)	If obstetricians have different propensities for abdominal delivery If lack of comparative information impedes obstetrician decision making regarding optimal practice style	If obstetricians function primarily on a "target income" basis If obstetricians perform procedure to avoid consequent legal ramifications
Availability of hospital resources (administrative decision making)	If lack of anesthesia services on weekends or nights influences procedure rates If only one obstetrician is in practice (e.g., in a rural hospital) and service availability is limited	If administrators encourage frequent performance of procedure for bottom line considerations
Patient/societal preference (patient decision)	If patients prefer a "quick fix" for labor pains	If higher rates are among educated "older" women who expect perfect baby If society is fascinated by surgery and high-tech medicine If society is expected to continue paying for higher cost alternative patient management methods
State of medical knowledge and training	If younger obstetricians are not trained adequately in management of complications If extensive use of internal monitors increases the probability of correct diagnosis of failure to progress	

The most common of all indicators, the one most extensively reported nationally, is mortality. The assumption in describing and analyzing in-hospital mortality rates is that the incidence of mortality is substantially associated with the capability of a hospital to provide quality health care. The annual Medicare mortality reports of the Health Care Financing Administration (HCFA) and other, similar analyses have raised concerns about this assumption, however. Is the mortality rate indeed a valid representation of a hospital's performance? People have questioned the conceptual validity of that indicator, but not necessarily the construction of the indicator.

Questions have also arisen within the health care research community about patient satisfaction as an indicator of the quality of the services received. Time and time again, reports of a high level of patient satisfaction have appeared to be unrelated to the technical quality of care. Researchers have begun to wonder if aggregate measures of patient satisfaction have the desired validity as an indicator.

It is not necessary for validity to be absolute. There is no need to make the "perfect" the enemy of the good. Validity comes in different shades, and, depending on the questions asked, an indicator of less than perfect validity can still be useful. Of course, the goal in defining indicators should always be to identify those that provide the highest and most consistent validity, however. Because it is inappropriate to use a measure that is inherently not valid, the validity of the indicator must be assessed before any other characteristic.

The Reliability Question

Much less problematic in nature than the validity of an indicator, the reliability of an indicator pertains to its level of measurement error. Simply put, the less error in the measurement, the better the indicator. Therefore, decisions about the reliability of a measure concern the mechanics of the construction of that measure or indicator.

The level of error in the measurement of a rate determines the extent to which that rate is reliable, just as the level of static, or noise, in the signal from a television station determines the reliability of the television reception. The extent to which error occurs in calculating a rate determines the extent to which that calculation is reliable (Table 3–2). Thus, the reliability is as much a function of the skill used in the rate construction as it is of the quality and availability of the data.

Many believe that a rate can be valid, but not reliable. In fact, this question is at the heart of a continuing debate on health care indicators. It is a major issue of contention, especially when indicator rates obtained in a number of different hospitals are to be compared.

No one can disagree that many health care indicators are conceptually valid tools or measures that can shed light on issues of performance and, perhaps, quality of care. Readmission rates, infection rates, complication rates, length of waiting time in the emergency department, and returns to

Table 3–2 Caesarean Section Rate As an Indicator of Quality Care: Reliable?

	Yes	Questionable
Numerator data	If all cases of Caesarean section are documented and retrievable from the hospital's information system (manual or computerized) If definition of numerator data elements has not changed over time (perhaps a root issue for comparable analysis of Caesarean section rates)	
Denominator data	If records of all delivered patients are retrievable from the hospital information system (manual or computerized)	
Epidemiological adjustments of data	If data about patient age (or at least age group) and parity are available	If data about vaginal births after previous Caesarean section (VBACS) are not collected and not readily accessible

the operating room are all measures that *may* reflect a hospital's performance. It is rare, however, to find overwhelming consensus among researchers that data for those indicators are or can be captured *uniformly* across large numbers of hospitals. This leads to questions about the extent to which interhospital comparisons of indicator rates are appropriate and justified.

Indicators can also be reliable, but not valid. That is, perhaps, a much more serious problem in indicator development than is the previous problem. For example, an in-hospital mortality rate, a commonly used indicator, may be only marginally associated with the goodness of a hospital's performance—even though the mortality data can be captured reliably and trends over time can be identified. Similarly, indicators can be reliable and reveal little about the appropriateness of care. A hospital may have highly skilled orthopedic surgeons and surgical staff who perform unnecessary, yet successful, laminectomies or hip replacements. In that situation, the mortality and/or postsurgical wound infection rates may be low, but there are questions about the appropriateness of the care. In other words, is it more important that the wrong things may have been done very well or that the right things have been done? That is a dilemma that each person interested in the validity (the appropriateness) and the reliability (the effectiveness) of health care indicators has to face.

RATE CONSTRUCTION

The Struggle

Many people struggle with the basic concepts behind a rate. Some of the most frequent, erroneous statements heard from health care professionals who are learning about tools and methods of performance assessment include

- "A rate is a description of how well we do in our hospital."
- "A rate shows how many procedures we performed this year."
- "A rate is a number that our finance department uses a lot."
- "I know what a percentage is, but not a rate."

Elements of a Rate

A rate is a statistic that describes how much has happened, to whom, and when. "How much" is the frequency of that happening. For example, there may have been 25 Caesarean sections at a hospital. Knowing how many abdominal deliveries were performed does not provide the rate of Caesarean sections, however. It is necessary to know 25 of how many deliveries.

The second element of a rate, the "to whom," is the group most likely to have experienced the observed happening. In this case, it is the group of all women who delivered at the hospital. In epidemiological terminology, this group is the population at risk. The term *population* refers not only to the demographic entity that health planners are interested in studying, but also to the group under study. In this example, 100 women delivered at the hospital. There is still no rate, however, since the time period within which the Caesarean sections took place remains unknown.

The 25 Caesarean sections performed for the population of 100 women may represent the delivery room activities for the month of July. Now it is possible to calculate a rate. The formula of a rate is as follows:

$$\text{Rate} = \frac{\text{How much happened}}{\text{To whom}} \quad \text{Within what time period}$$

This is the simplest way to remember the construction of a rate. The numerator of the rate is "how much," the denominator is "to whom," and the time period represents the "when." In the example on Caesarean section, the rate becomes:

$$\text{Caesarean section rate} = \frac{25 \text{ Women of}}{100 \text{ Women delivering}} \quad \text{in July 1993}$$

The rate is, therefore, 25 percent for the month of July.

Another example demonstrates the importance of the proper display of a rate. An informative letter from the county department of health states that, in 1992, the fluoride content of the drinking water in the county is less than 1 part per million (ppm), whereas the figure for the neighboring county is 2 ppm. Is this a rate? Indeed, in 1992, fluoride "happened" to "water" with a frequency of 1 fluoride unit to 1 million units of water, in the county. That is a rate, although it is not expressed as a percentage. Because a percentage simply expresses the measurement in the numerator "per cent" (i.e., per 100) of the denominator, the 1 per 1,000,000 (ppm) translates to 0.0001 per 100. Surely, 1 ppm is easier to understand and remember.

The unit of measurement is obviously important. As a rule of thumb, whole numbers are easier to use than are fractions—1 per 1,000 is a better expression than 0.1 percent.

ERRORS OF CONSTRUCTION AND INTERPRETATION

The analysis of error is a fundamental statistical mandate. Those who evaluate performance may find it difficult to accept measurement errors. It is essential, however, to know what type of error is inherent to the measurement of a phenomenon (e.g., mortality rate, delays in getting laboratory reports, error in x-ray interpretation rates).

Most Common Errors

Sometimes, knowing what a rate or indicator is may be more challenging than understanding what it is not. The following are some of the common types of confusion when detecting and interpreting errors of rate construction.

- "Frequency is a rate." No. Frequency is the *numerator* of a rate.
- "The denominator of the rate should be all discharges." No. The denominator should be the population at risk. For example, the denominator in the rate of prostate cancer may be all men above 50 years of age; that in the rate of tonsillectomy may be all patients between the ages of 2 and 18 years; and that in the rate of total hip replacement for osteoarthritis may be all patients above the age of 55 years who were radiologically confirmed to have osteoarthritis.
- "All rates should be percentages." No. As shown earlier, low-frequency events—also called rare events—need a larger population base for detection. A rate can be expressed in any unit of magnitude; the importance is to display that rate in a logically acceptable format.
- "A high rate is always bad." No. A high rate is high; that is all. An arbiter or judge subsequently assigns a value to the observed value-free statistic.
- "A percentage is not a rate; it's a per cent." Clearly not. Rates can be expressed as percentages, as well as per 1,000 or per 10,000.
- "A rate indicates how well we are performing in our hospital." Not really. A rate often describes what staff are doing in the hospital. The evaluation of that rate is reserved to the educated, cautious, and patient interpreter of that value-free statistic.

Types of Errors

There are two types of errors: systematic errors and random errors. The underlying concepts are simple:

> A systematic error is an error that is built into the logic of the measurement. That is, it is always there, and it is predictable.
>
> A random error is a mistake in measurement that happens in a haphazard way and is mostly unpredictable.

Systematic Error

According to some studies of functional status and quality of life, married persons are healthier and happier than are their unmarried counterparts. It may be tempting to conclude that "marriage is good for you." It is possible to argue that a certain self-selection may exist before marriage, however. Persons who are successful in marriage may have predisposing character traits that account for the observed differences in health status. Studies that do not control for such confounding factors introduce a systematic error of measurement into investigations of marriage's effect on health status.

Clearly, a number of issues interplay throughout the two incidents described. Among those are issues of confounding variables, faulty study design, and selection bias. The resulting type of error, namely, the systematic error that underlies these measurement designs, influences the validity of a rate or indicator. Basically, the critique of such a measurement is that it measures the wrong thing. Such statements are heard when patient condition severity is not fully taken into account; when there is a selection bias in the patient mix between two hospitals whose mortality rates are being compared; and when, instead of discussing obstetricians' proclivity for Caesarean section, measures focus on patient preference or even the ambient payment system for health care.

Random Error

Much easier to detect than are systematic errors, a random error appears in measurement "swings" over time. That is, a blip appears on the trend curve if an error occurs in the collection of the data. There are a number of commonly encountered random error scenarios for indicators of care. For example, a sudden surge or decrease in the nosocomial infection rates over time may occur for a variety of reasons, including the following:

- Not all the necessary data were collected (numerator error).
- Not all patients at risk for infection were identified (denominator error).
- A new infection control professional was hired in that time period, and this person carries on surveillance differently.

In any of these circumstances, the measurement errors are random and can be rapidly corrected.

Random errors may also result from the use of multiple formulas to measure the same phenomenon. Ten hospitals may share their readmission rates, for example. In five of these hospitals, the information system is fully computerized, and all readmissions are documented in a timely manner. Four of the remaining five hospitals have a mix of manual and computerized data-gathering systems, and the fifth hospital relies totally on manual data gathering and tallying. When data for the first month are shared, one administrator asks the question, "Are we sure we are all collecting data the same way?" It is a rather logical inquiry, but not an easy one to answer; it is a question of data reliability. To be sure, special reliability assessment studies need to be ongoing. Present knowledge from a few such studies suggests a higher reliability (less random error or "noise" in measurement) among the first five hospitals and a noteworthy difference in data reliability between the first five and the last (manual data gathering) hospital's data.

Indicator Acceptability

As discussed earlier, the concepts of reliability and validity are not interchangeable. Measures, indexes, or indicators *must* be valid. Reliability, on the other hand, is a matter of degree. It is important to decide what degree of reliability is acceptable, however.

Two variants of measurement "noise" are of interest to hospitals that share and compare data. First is the question of how much noise exists for an individual measure at a single hospital. A trend analysis may identify sudden changes in rates, but ad hoc investigations are necessary. Second, when comparative analysis is performed, it is essential to know if all the hospitals are collecting these data with a comparable amount and type of error. Perhaps this question is one of "contrast reliability." Once these questions have been addressed, it is time to decide how much error is acceptable. This decision depends on the nature of the indicator and the type of analysis to be performed (Table 3–3). In some instance, 80 percent reliability may be adequate; in others, nothing less than 95 percent would do.

CONCLUSION

Indicators have always been used in epidemiology to study the frequency of disease occurrence in populations. They are the primary representations of the financial market's daily behavior; they constitute every baseball fan's main method of predicting a team's performance; and they

Table 3–3 Relationship between Types of Errors and Acceptability of a Measurement

	Systematic Error	*Random Error*
Validity*	Minimum to none	Not a primary issue
Reliability†	Not a primary issue	Low
An "acceptable" measure (e.g., indicator)‡	Minimum to none	Low

* Commonly heard argument when clinical indicator rates are shared with physicians. For example, if there is a systematic omission of high risk patients from a measurement, then the clinical validity of that measure is questionable.

† Usual source of discourse when two or more hospitals' data are compared and contrasted.

‡ The basic imperative for evaluating the goodness of a measure.

have been taught in every introductory sociology course. Health services researchers and policy makers are now rediscovering the potential usefulness of indicators. Reasons are varied. Perhaps it is because resources are limited; perhaps it is because the traditional belief that medical care can be provided only by disciples of Hippocrates is being challenged; or perhaps it is because the concepts of health and well-being are evolving. No matter what the web of influence, the request for better methods of measurement and reporting has become part of the health care profession's social accountability mandate. It is necessary to create not only better measures, but also better understandings of the functioning of these measures. In addition, appropriate assessment and accurate interpretation of the errors of these measures are critical.

Irrespective of the limitations of the measures, social inquisitiveness is developing—and that is good. It is important to keep in mind yet another merry saying, however:

"How many psychologists does it take to change a light bulb?"
"One. But the light bulb should be ready to be changed. . . ."

NOTE

1. V.A. Kazandjian, Indicators of Quality: Pointer Dogs in Disguise, *Journal of the American Medical Records Association* 62, no. 9 (1991):34–36.

» Chapter 4 «

Survey Methods for Quality Improvement Professionals

Michael A. Counte and Kristen H. Kjerulff

AS NOTED by many contemporary observers, we are truly in the midst of an information age. Because surveys provide much of the information that is needed in our information-oriented society, they are a pervasive part of everyday life. For example, surveys are used to monitor prevailing political perceptions and social attitudes (public opinion polls), assess levels of political preferences, uncover consumer sentiments regarding the purchase and use of goods and services (market research), measure levels of unemployment, track the frequency of criminal events, determine the costs of health services and related expenditures, and describe the amount of foodstuffs that are produced in agricultural areas. These are just a few of the major applications of surveys in everyday life.[1]

Most authors agree that survey research is characterized by three attributes.[2-4] First, the purpose of survey research is to provide quantitative estimates or descriptions of select phenomena. This clearly distinguishes survey research from open-ended interviews that seek qualitative, rather than quantitative, responses. Second, survey researchers ask people questions in person, over the telephone, or via the mail. In other words, there is little, if any, direct observation of behavior involved. Finally, and perhaps most important, survey research focuses on only a fraction of the population or a sample. Thus, it does not comprise census methods that enumerate the characteristics of an entire population. Advances in sampling methods over the last 40 years have been remarkable. In fact, it is because of such extraordinary advances in sampling strategies that contemporary surveys can focus on very small samples and yet produce findings that are truly generalizable to the larger population under study.

Contemporary survey research has a variety of purposes. Health care organizations commonly use surveys to document the existence of relevant phenomena and to uncover underlying factors and trends. The simplest use is to provide a *description* of select variables. For example, a hospital researcher may be interested in determining the number of patients discharged from an inpatient facility who return to the facility if they subsequently need additional services. The rate may then be compared to those of other, similar institutions (benchmarking). An investigator who finds that the return rate observed for his or her hospital is low may once again conduct a survey of another group of discharged patients. This time, the goal is *explanation* rather than description. The primary issue in the new survey is to discover why the return rate at the hospital under study is low. It is, of course, difficult to anticipate the results of such a project, but several factors may explain a lower return rate, including the types of patients served by the facility, its location, or attributes of the treatment provided. Unfortunately, even after the second survey, there still may be considerable uncertainty regarding the reasons for the facility's low return rate. In this case, additional in-depth surveys may be conducted for *exploration* of the problem under study.

SURVEY RESEARCH METHODS

Research Hypotheses and Questions

The first and probably the most important step in planning a survey is to delineate the research hypotheses and/or questions that the survey study is intended to answer. Generally, it takes considerable effort to construct specific and focused research hypotheses or questions that reflect the objectives of the study, specify the variables to be measured, and identify the population under investigation. For example, "Private patients visiting physicians in the obstetrics and gynecology department at Hospital X will be less satisfied with office waiting time than are clinic patients" is a specific and focused hypothesis. It reflects the objective of the survey, which is to discover why the number of private obstetrics and gynecology patients is dwindling at Hospital X. The variable under investigation is satisfaction with office waiting time. Two study populations are identified in this hypothesis: private and clinic patients who are visiting the obstetrics and gynecology department at Hospital X. Sometimes, a survey researcher has no particular hypothesis in mind, but prefers to specify research questions to be addressed by the survey. The corresponding research question for this example is, To what extent do private patients visiting the obstetrics and gynecology department at Hospital X differ from clinic patients in satisfaction with waiting time? The development of a list of specific, focused, and feasible hypotheses or research questions as part of the planning process for a survey study helps to ensure that the objectives of the study are achieved.

Survey Research Designs

There are many potential designs for survey studies. The choice of design depends on the research question to be addressed, the funding available for the study, and the time period in which the study must be completed. The most common survey designs are cross-sectional, longitudinal, experimental, and quasi-experimental.

Cross-Sectional Survey Studies

Researchers conducting cross-sectional studies collect data at one point in time. They may collect data on one population of interest, or they may compare data on specific subgroups. A researcher may, for example, wish to survey all patients who attend the family practice clinic during the month of October in order to compare the care received by specific subgroups, such as Medicaid and private insurance patients. Cross-sectional studies have the potential advantage of yielding results relatively quickly, because data are collected during a finite time period.

Longitudinal Survey Studies

In longitudinal survey studies, data collection takes place at more than one point in time. Researchers may be studying the natural history of some particular phenomenon. For example, the obstetrics and gynecology department staff at Hospital X may wish to survey their private patients at regular intervals in order to identify at what point and why some of these patients decide to go elsewhere for their care. Longitudinal studies may also be used to examine the impact of a particular event or change in practice. If, for example, the staff at Hospital X want to evaluate the impact of a new appointment-scheduling system, they may interview patients before and after the implementation of the new system in order to assess changes in patient satisfaction with the appointment-scheduling process.

Longitudinal studies have the advantage of allowing the investigation of potential cause-and-effect relationships. An increase in patient satisfaction with the appointment-scheduling process after implementation of a new system may be due to a new system; however, some other change, such as the addition of a new and more pleasant staff person in the scheduling office, may be responsible for the increased satisfaction with appointment scheduling. In longitudinal studies without control, there is a risk that unmeasured variables are actually causing an apparent relationship between an event and an outcome of interest.

Experimental Survey Studies

The most basic experimental survey study is one in which the researcher randomly assigns individuals to one of two groups, surveys both groups on the factors of interest, exposes one group to some particular treatment or event, and then surveys both groups again on the factors of interest. To conduct an experimental study of the effect of a new appointment-scheduling system, for example, an investigator may randomly assign patients calling for appointments to one of two groups, either the intervention group or the control group; survey both groups as to their satisfaction with the appointment-making process; expose the intervention group to the new appointment-making system the next time they call for an appointment, while using the regular appointment-making procedures when the control group call; and again survey patients as to their satisfaction with the appointment-making process. If the two groups were equivalent in satisfaction prior to the intervention, but the group that received the intervention (the new appointment-scheduling system) becomes significantly more satisfied (or less satisfied) than the control group after the intervention, the postintervention difference is likely to be due to the intervention under investigation.

Experimental survey studies have the advantage of allowing an examination of cause-and-effect relationships with more certainty that the specific cause has been identified than longitudinal studies. In many instances, however, it is neither feasible nor ethical to expose only some people to the intervention of interest.

Quasi-Experimental Survey Studies

Occasionally, a researcher wants to conduct an experimental study, but for some reason cannot use random assignment to designate intervention and control groups. It may be necessary to put people who are willing to receive a particular intervention in the intervention group and everyone who is not willing to receive the intervention in the control group. Patients from one clinic may form the intervention group; patients from another clinic, the control group. Whatever the method of assignment, it is important that the factors of interest be measured in both groups prior to the intervention to ensure that the two groups are equivalent prior to the intervention. This type of study uses a quasi-experimental design, a good choice if random assignment to groups is not possible, feasible, or ethical.

Measurement of Variables of Interest

Well developed research hypotheses or questions guide the researcher in selecting the variables to be measured. The next step is to decide how best to measure each variable, that is, each variable's operational definition. A specific variable may have many potential operational definitions. The definitions chosen for key variables largely determine the extent to which the results of the study are accurate, precise, and unbiased. One researcher may choose to measure patient satisfaction with medical care with a single question, while another uses a 30-item inventory. These two approaches can yield completely different results. Thus, a researcher must select the operational definitions of variables very carefully.

It is often best to begin the process of deciding how to measure specific variables by conducting a literature review and gathering measures used by other researchers. The more thoroughly this search is done, the more likely the survey questions will be good. For example, a very thorough literature search may produce 50 different inventories designed to measure patient satisfaction with medical care. Many of these inventories measure aspects of medical care that are not pertinent to a particular study for one reason or another, or they may not fit in with the study's definition of satisfaction with medical care. Certain inventories may be too simplistic; some may be too long or complex. Some inventories may have questions that appear confusing, contain ambiguous terminology, or use poor grammar. The process of thoroughly reviewing the literature and carefully examining other data collection instruments, however, facilitates the development of operational definitions for key variables that appropriately address the research questions and objectives in the study.

Instrument Credentialing (Validity and Reliability)

Through instrument credentialing, researchers attempt to establish the worth of a data collection instrument. Generally, its worth depends on the extent to which it consistently and accurately measures the concept of interest. Measurement error arises from a variety of sources, including ambiguous or poorly worded directions and questions, sloppy and inconsistent instrument administration procedures, the use of language that some percentage of the respondents do not understand, and respondent guessing or faking of answers.

The best guide for the identification of many sources of measurement error is common sense. For example, questions that require people to rely on long-term memory (e.g., How many times have you been to see a dentist in the past 10 years?) will probably yield poorer data than do other types of questions (e.g., How many times have you been to see a dentist in the past year?). In addition, many questions elicit some knowingly untrue answers, because some people try to appear "better" than they really are. Therefore, they will claim to have been to the dentist in the past year when they have not. Occasionally, researchers verify answers to specific questions for a random subset of the respondents in order to examine the extent to which these questions elicit accurate answers. A researcher may check dental records for a subset of people asked about dentist visits in the previous year, for example. This type of instrument-credentialing process is important when decisions are being made on the basis of survey results.

Data collection instruments are often evaluated on the basis of their validity, the accuracy with which a particular concept is measured, and their reliability, the consistency with which an instrument measures the concept. The assessment of the validity of a data collection instrument is usually more difficult than the assessment of its reliability.[5] Many concepts that are important determinants of health-related decisions are primarily subjective in nature. The validity of a data collection instrument is usually measured by examining the extent to which it predicts specific, expected behaviors. For example, patients who are very satisfied with the care that they are receiving in an obstetrics and gynecology department are likely to recommend that department to their friends and to return to the department if they need additional care. If scores on a satisfaction instrument are not related to the expected behaviors, the instrument may not be valid for these purposes.

If the concept to be measured is assumed to be relatively stable (e.g., intelligence or manual dexterity), then an instrument designed to measure it should yield relatively similar scores each time it is administered. This is referred to as test-retest reliability.[6] If a data collection instrument is composed of multiple items that are summed to create a total score (e.g., the Scholastic Aptitude Test), then each of the items should measure the same concept of interest. This is referred to as internal consistency reliability. Authors of data collection instruments can often provide information concerning the reliability of their instruments to help researchers decide which one is the best for their purposes.

Methods of Data Collection

Once the data collection instruments have been developed or chosen for a particular study, the researcher must decide how to collect the data. Survey studies are usually conducted by telephone, in-person interviews, or mail. There are distinct advantages and disadvantages to each of these modes of administration.

Telephone Surveys

There are several advantages to collecting data by telephone interviews. Not only is telephone interviewing relatively easy and inexpensive, but also results can be obtained quickly, particularly if responses are entered directly into a computer as questions are being asked. Many survey research centers and other institutions now have computer-assisted telephone interviewing (CATI) services that can generate survey results quite efficiently. In addition, it is relatively easy to obtain or generate randomly sampled telephone numbers.[7] Study samples obtained via random sampling procedures are often preferable to samples obtained in other ways.

One major disadvantage of telephone interviewing is that only people with telephones will be interviewed. In some low-income urban and rural neighborhoods, less than 80 percent of households have telephones. If it is important for research purposes to interview low-income individuals, then telephone interviewing may not be appropriate. Another disadvantage of telephone interview studies is that the samples tend to contain a disproportionate number of those who are elderly and those who are not employed outside the home, because these people are more likely to be home when the interviewers call. In addition, lengthy or complicated surveys, as well as surveys concerning intimate or personal topics, are difficult to conduct by telephone. Respondents tend to grow impatient with such surveys and hang up before the interview has been completed.

In-Person Surveys

It is easier to conduct lengthy, complicated, and personal interviews in person than by telephone or mail. An interpersonally skilled interviewer can establish a pleasant interpersonal dynamic that keeps the respondent interested and happy to cooperate, even for quite lengthy surveys. Furthermore, such an interviewer can usually determine by verbal and nonverbal cues when a respondent does not quite understand a survey question and can be trained to provide additional explanations or examples as needed. When a researcher intends to survey people across socioeconomic levels, it is often preferable to conduct in-person interviews in order to be certain that all respondents, even those who cannot read and do not have telephones, understand and answer the questions. Studies concerning topics that are quite personal in nature can also be successfully conducted by in-person interviewing, as long as the interviewers are trained to convey the reason for such questions and to ask personal questions in a very matter-of-fact manner.

The primary disadvantage of in-person interviewing is that it is expensive. In-person interviews are often conducted in respondents' homes, and interviewers are usually paid for the time it takes to arrange an interview, travel to and from the respondent's home, and conduct the interview.

Mail Surveys

Usually, mail surveys are considerably less expensive than are telephone or in-person surveys. In addition, mail surveys can reach thousands of individuals in a relatively short time period. If a study requires surveying a large number of people in a short period of time, then it is probably best to conduct a mail survey. Mail surveys generally elicit fairly low response rates in comparison to in-person or telephone surveys, however. Many people automatically throw mail surveys into the trash. In addition, many people have difficulty answering written survey questions; they misunderstand the survey directions, skipping items, skipping pages in the questionnaire, and providing illogical or inconsistent answers. Successful mail surveys are usually brief, quite simple, and relevant to a topic that is of interest to the respondents.

Sample Size Estimation

Surveys are not usually administered to the entire population of interest, because such a survey is too expensive and/or time-consuming. Instead, some subset of individuals is surveyed. Before conducting a survey, the researchers must determine if the sample size is large enough to allow an adequate test of their hypothesis. If, for example, they want to compare satisfaction with care for private versus clinic patients in the obstetrics and gynecology department at Hospital X, they would survey ten patients in each group. It is quite possible, however, that this would be far too few subjects to test the hypothesis adequately. On the other hand, 1,000 subjects in each group may be more than is necessary.

A researcher who is statistically testing a difference between two groups may mistakenly reject the null hypothesis (i.e., there is no difference between the groups) and conclude that the two groups differ when they do not. This is called a Type I error. The probability of making a Type I error, also known as alpha, is often predetermined to be either .05 or .01. Researchers generally wish to make the probability of making a Type I error rather small. It is also possible, however, to fail to reject the null hypothesis of no difference between the groups when there really is a difference. This mistake is called a Type II error. The probability of making this mistake is called beta, and the probability of *not* making this mistake is called power.

The power of a statistical test of a hypothesis is determined, in part, by the sample size.

If, for example, a researcher expects that approximately 90 percent of the clinic patients and 80 percent of the private patients will agree with the statement, I plan to continue to receive care for my gynecological and obstetrical needs at Hospital X for the foreseeable future, then a two-tailed hypothesis test with an alpha level of .05 would require a sample size of 220 subjects in each group in order to have an 80 percent chance (power) of rejecting the null hypothesis.[8] If the researcher wishes to have a 90 percent chance of rejecting the null hypothesis, then a sample size of 280 subjects in each group is necessary. Sample size tables are available.[9]

The Sampling Plan

Once the sample size needed has been determined, a plan for obtaining subjects (sampling plan) must be developed. Ideally, the sample will represent the larger population. There are many potential sampling plans.[10] In the study example described earlier, the researcher may ask private and clinic patients who come to the obstetrics and gynecology department at Hospital X if they are willing to participate in the study, continuing this process until the desired sample size has been obtained. This constitutes a nonprobability sample. On the other hand, the researcher may randomly sample among lists of private and clinic patients who have seen physicians at the obstetrics and gynecology department at Hospital X in the past year. This technique produces a probability sample.

Probability samples are usually preferable to nonprobability samples, because the probability that any one individual will be chosen and asked to participate in the study is known in advance and because each individual has an equal chance of being chosen. As a result, the obtained sample is truly representative of the larger population. It is frequently not feasible to use a probability sample design, however. In the case of the survey of private and clinic patients from the obstetrics and gynecology department at Hospital X, for example, there may be no easy way to generate a list of all patients who have seen physicians in the previous year. Thus, a researcher may have no choice but to obtain potential subjects by using some type of nonprobability sampling plan.

The Pilot Study

Once the data collection instrument has been developed and the sampling plan completed, a small-scale version of the planned study, a pilot study, is conducted. This is an absolutely essential step in the development of a respectable survey study. Sometimes, researchers conduct multiple pilot studies until they are certain that everything is functioning smoothly. A well conducted pilot study addresses many issues:

- Do potential participants understand each of the survey questions?
- Do the questions elicit an appropriate level of variation?
- Are the survey directions clear and understandable?
- Is the survey too long or time-consuming?
- Is every question essential?
- Does the structure of the questionnaire lend itself to skipped items?
- Are there questions that people cannot answer with complete certainty?
- Do the procedures planned for obtaining subjects work as planned?
- Does the intended data collection procedure (i.e., telephone, in-person, or mail survey) work as planned?
- Are there additional concepts that should be measured and questions that should be added?
- Is the questionnaire structured so that it is easy to enter answers into the computer and analyze the data?
- Will the study cost about as much as planned, or are there costs that were not anticipated?

A well conducted pilot study should uncover any potential flaws within the survey instrument and data collection procedures.[11] It is substantially preferable to uncover flaws during the pilot study than after the actual study has begun.

SURVEY RESEARCH APPLICATIONS

Analysis of Customer Satisfaction

There is little question that the measurement of customer (previously patient) satisfaction represents a long-standing application of survey research methods within various types of health care organizations—hospitals, medical practices, and managed care facilities (e.g., staff model health maintenance organizations).[12] Generally, customer satisfaction refers to a person's attitudes toward a product or service. Dissatisfaction with a facility seems to result largely from a discrepancy between a customer's expectations and the perceived quality of his or her actual experience while in treatment. The facility's emphasis is on trying to identify the reasons for such a discrepancy and ways to remedy the situation. A desired outcome, of course, is to have the quality of patients' experiences substantially exceed their expectations of the service. In addition to measuring the level of customer satisfaction, many facilities are now also questioning patients about the likelihood that they will return to the facility where they have received treatment and whether they will recommend the facility to any friends or neighbors who require similar services.

Interest in the issue of customer satisfaction has increased significantly in the last 10 years.[13,14] There are many reasons for this augmented interest. First, customer satisfaction is now known to be a key measure of organization and product line effectiveness or quality within health care organizations.[15] Second, in an age of increased emphasis on accountability, customer satisfaction represents a prime indicator of an organization's responsiveness to patients, their families, and health care purchasers. Third, the widespread introduction of guest relations and continuous quality improvement programs highlights the need to enhance customer satisfaction.[16] Fourth, increased competition among health care organizations in various sectors of the health care industry has increased the salience of customer satisfaction. Finally, customer satisfaction is now recognized as an important impact variable that needs to be monitored when new programs and technologies that affect the quality of health care services are introduced.[17]

As a result of such industry trends, health services researchers have turned their attention toward the methodological issues involved in attempts to assess patient satisfaction and strategies for improvement.[18,19] Also, investigators have increased their focus on specific types of problems that significantly affect patients' overall satisfaction and the prevalence of such difficulties in large samples.[20,21] Although largely focused on hospital inpatients, these studies are important to quality assurance professionals, because the researchers have drawn their samples from a diverse group of individuals who have been discharged from various types of facilities. Thus, the results are more likely generalizable to hospital patients in general than are findings from isolated single-site studies. Such projects have consistently shown that the interpersonal aspects of care (e.g., courtesy, sensitivity, concern) have a major effect on the quality of care from the patient's perspective. The perceived quality of a patient's interaction with both clinical and support staff influences overall patient satisfaction much more heavily than do hotel services such as the quality of the food, room temperature, or noise level.

Despite the numerous studies of patient satisfaction that have been reported and the relatively recent appearance of "how to" texts,[22] a number of issues continue to require the attention of health care managers interested in quality improvement. These concerns are especially important now, because most health care organizations are currently assessing patient satisfaction in some manner. These questions basically can be grouped into five areas:

1. appropriate personnel to perform the assessment. Hospital managers need to be aware that using in-house staff and hiring an external consulting firm with expertise in patient surveys both have advantages and drawbacks in conducting patient surveys. Even emerging systems of facilities should be sensitive to this issue.
2. performance benchmarks. The need to obtain comparative patient satisfaction results is fairly self-evident. One major advantage of using the services of a survey research firm is that comparative data should be more readily available. Benchmark results can often be obtained from other facilities through a trade, without subscribing to a service, however.
3. patient satisfaction and other quality indicators. Patient ratings of the quality of services that they receive are oriented primarily toward what has historically been called the "art of medicine." This is not surprising, because it is clearly difficult for patients to evaluate the technical aspects of their care. Nevertheless, measures of technical quality are frequently monitored in health care organizations, and their relation to patient satisfaction can be readily studied. After all, both patient and expert ratings are deemed important quality indicators.
4. satisfaction with dimensions of care vs. overall ratings. Although overall satisfaction is an important global indicator of the quality of care from the patient's perspective, a health care organization should be able to differentiate levels of satisfaction with different aspects of the care episode. As noted earlier, there is an increasing agreement concerning the dimensions that need to be studied, as well as an emerging consensus that certain areas (e.g., interpersonal interaction with clinicians and support staff) are particularly important.
5. methodological issues. Among the numerous methodological issues that must be considered during the survey design and data collection process are the type of sampling plan that will be used and the way in which data will be collected. The major goals should be to assemble samples that are very typical if not representative of the patient population and to achieve a response rate that is similar to those achieved in comparable settings. A survey service or a local

expert in survey research may be able to help with the resolution of these issues, if necessary.

Survey Data and Market Research

Although market research has long been an important function in business organizations, its growth in the health care industry is of much more recent origin.[23] There are three important reasons that health care organizations have recently become involved in sustained market research.[24] First, continual cost increases have strongly encouraged *competition* within all segments of the health care field. Second, *consolidation* of providers into integrated delivery systems has continued. Finally, *consumerism* has emerged as a major force in health care. Such pressures, which have increased in intensity only during the last 20 years, have prompted health care organizations to act more like business corporations and adopt a market-based planning approach.[25] Such an approach dictates that management base strategic planning decisions on data (rather than on the beliefs or perceptions of corporate executives).

Market research includes a wide range of activities, such as advertising studies, forecasting projects, corporate responsibility initiatives, analyses of new products, and studies of sales.[26] Surveys are very frequently used in market research both within and outside health care organizations. For example, a large facility may conduct periodic surveys of its medical staff in order to increase their linkage to the facility. Survey responses allow management to better understand the needs and wants of the medical staff. In addition to such internal survey assessments, health care organizations frequently conduct or sponsor community-based consumer studies in an effort to measure current perceptions of their services and perhaps also to project future expectations and needs.[27]

Although the survey method is a powerful tool in market research activities, it is important to keep in mind several concerns:

- timing of market research. It is fairly evident that market research is a powerful proactive tool that can help an organization make effective strategic decisions in the midst of an ever-changing marketplace. Surveys can also be useful after a program has begun, however, in order to assess its level of market penetration and perhaps to identify obstacles to its future success.
- use of primary vs. secondary data. Surveys that provide useful market research data need not be sponsored by any particular facility. Secondary data, which are collected by other organizations (e.g., gov-

ernment units), can be very useful to a facility in its planning efforts if the right questions have been asked, the sample size is acceptable, and the data are of relatively recent origin. Thus, before the start of a primary data collection project, it is wise to determine whether any relevant secondary data are available.
- method of data collection.

Surveys and Human Resources Management

The human resources management unit of a health care organization coordinates the selecting, hiring, training, evaluating, and retaining of employees. In an era increasingly characterized by shortages of qualified personnel in certain areas, pressures for greater efficiency and effectiveness (exemplified by current down-sizing in the hospital sector), and greater diversity in the workforce, human resources management continues to be very important.

Survey tools can be particularly useful in four areas of human resources management. First, community-based surveys permit continual monitoring of the attributes, expectations, and needs of persons living in the local area. Second, surveys of facility employees can indicate their reactions to in-house procedures that govern staff recruitment, training, and performance evaluation. Third, a survey is an ideal instrument to monitor factors such as employee work motivation and intent to leave. Finally, employee surveys are very useful in delineating the employee impacts of organizational changes.

Surveys in Technology Assessment

Modern health care organizations are almost continually implementing new clinical and managerial technologies. Recent managerial innovations in the health care field include integrated management information systems, standards-based cost accounting procedures, restructuring and re-engineering of work processes, and quality improvement programs. The development/acquisition and implementation costs of such new technologies can be very substantial for the host institution, particularly in view of the tighter constraints on reimbursement for services.

Survey methods can be very useful in technology assessment, because the introduction of a managerial innovation within an organization can be considered an experimental treatment. Thus, by studying an organization's employees before and after the implementation of an innovation, an investigator can use a quasi-experimental design to identify the innovation's effects more accurately. Although this approach does not include a control group as is the case in true experiments, it is still preferable to post-treatment only analysis.

Counte and colleagues recently used a survey approach to characterize work-related attitudes and opinions before and after the introduction of a large-scale total quality management (TQM) initiative.[28] The results of their 2-year study provided strong support for the general contentions of TQM proponents that employee involvement in such programs results in a range of positive outcomes, including higher job satisfaction levels, more favorable work-related perceptions, and a more positive view of the organization's climate. These effects were significant, even when sociodemographic variables were controlled.

CONCLUSION

Survey methods are very useful in the evaluation process, irrespective of the particular question under study. Surveys are as useful to health care managers as are any other tools that they employ, such as forecasting techniques or other decision support tools. Therefore, it is difficult to understand why managers of health care organizations do not use surveys more often. Several possible explanations are available. There may be a mystique that surrounds survey research, and managers may feel that surveys are invalid and unreliable unless survey experts have produced them. Another plausible explanation is that survey data are too difficult to analyze.

There is little question that contemporary management within health care organizations is heavily data-driven. Managers need data to formulate decisions. The survey method is very helpful in efforts to address questions such as

- How satisfied are our patients?
- What affects the likelihood of patients returning to our facility when the need arises?
- What do our employees expect from their work?
- What effects do major organizational interventions have on employees?

Some may contend that it is impossible to obtain answers to such questions, because they are too dependent on the subjective responses of individuals. The development of contemporary survey research methods in the last 50 years provides substantial evidence that, although phenomena such as attitudes, beliefs, and behavioral intents may be difficult to measure, techniques now exist to simplify the process. Contemporary managers need to know when and how to use survey methods and how to use the findings that are obtained to improve the quality of the decisions that they make.

NOTES

1. F.J. Fowler, Jr., *Survey Research Methods,* Vol. 1, Applied Social Research Methods Series (Beverly Hills, Calif.: Sage, 1984).
2. L.A. Aday, *Designing and Conducting Health Surveys: A Comprehensive Guide* (San Francisco: Jossey-Bass, 1989).
3. E.R. Babbie, *Survey Research Methods* (Belmont, Calif.: Wadsworth, 1973).
4. P.J. Lavrakas, *Telephone Survey Methods,* Vol. 7, Applied Social Research Methods Series (Beverly Hills, Calif.: Sage, 1986).
5. E.G. Carmines and R.A. Zeller, *Reliability and Validity* (Beverly Hills, Calif.: Sage, 1979).
6. Ibid.
7. J. Waksberg, Sampling Methods for Random Digit Dialing, *Journal of the American Statistical Association* 73 (1978):40–46.
8. M.W. Lipsey, *Design Sensitivity* (Beverly Hills, Calif.: Sage, 1990).
9. Ibid.
10. G. Kalton, *Introduction to Survey Sampling* (Beverly Hills, Calif.: Sage, 1983).
11. Aday, *Designing and Conducting Health Surveys.*
12. P.D. Cleary and B.J. McNeil, Patient Satisfaction As an Indicator of Quality Care, *Inquiry* 25 (1988):36.
13. J. Graham, Quality Gets a Closer Look, *Modern Healthcare* 17 (1987):20–31.
14. I. Press et al., Patient Satisfaction: Where Does It Fit in the Quality Picture? *Trustee* 45 (1992):8–10.
15. G. Glandon et al., Assessing Your TQM Program's Impact, *Strategies for Healthcare Excellence* 6 (1993):7–12.
16. C.W. Nelson, Patient Satisfaction Surveys: An Opportunity for Total Quality Improvement, *Hospital & Health Services Administration* 35 (1990):409–427.
17. M.E. Rindler, Back to the Patients: Process vs Outcome for Hospital Managers, *Hospital & Health Services Administration* 29 (1984):15–22.
18. K. French, Methodological Considerations in Hospital Patient Opinion Surveys, *International Journal of Nursing Research* 18 (1981):7–32.
19. P.L. Stamps, Measuring Patient Satisfaction, *Medical Group Management* 2 (1984):36–44.
20. I. Press et al., Satisfied Patients Can Spell Financial Well-Being, *Healthcare Financial Management* 45 (1991):34–42.
21. H.R. Rubin et al., The PJHQ Questionnaire: Exploratory Factor Analysis and Empirical Scale Construction, *Medical Care* 28 (1990):S22–S28.

22. See, for example, S. Strasser and R.M. Davis, *Measuring Patient Satisfaction for Improved Patient Services* (Ann Arbor: Health Administration Press, 1991).
23. P. Kotler and R.N. Clarke, *Marketing for Health Care Organizations* (Englewood Cliffs, N.J.: Prentice-Hall, 1987).
24. P.H. Keckley, *Market Research Handbook for Health Care Professionals* (Chicago: American Hospital Publishing, 1988).
25. S.G. Hillestad and E.N. Berkowitz, *Health Care Marketing Plans: From Strategy to Action*, 2nd ed. (Gaithersburg, Md.: Aspen Publishers, 1991).
26. Kotler and Clarke, *Marketing for Health Care Organizations*.
27. P.D. Cooper, ed. *Health Care Marketing: Issues and Trends* (Rockville, Md.: Royal Tumbridge Wells, 1985).
28. M.A. Counte et al., Total Quality Management in a Health Care Organization: How Are Employees Affected? *Hospital & Health Services Administration* 37 (1992):503–518.

» Chapter 5 «

Designing Quality Management Programs for Today and Tomorrow

Joann Genovich-Richards

HEALTH care services, like other major societal institutions, continue to change and evolve. Whatever the changes at the societal and health care industry levels, a specific health care institution's organizational design is one of the areas in which it can exercise choice. Shortell and associates noted:

> Hospital efficiency and quality of care can each be viewed as a function of the hospital's external environment, technology, and certain internal organization design variables. Relatively speaking, a hospital has little control in the short run over its external environment or technology. However, it can exert considerable control over the internal design variables.[1]

Organizational design (or structure) affects performance. Pfeffer defined organizational design as "the process of grouping activities, roles, or positions in the organization to coordinate effectively the interdependencies that exist."[2] Therefore, design is more than the organizational chart that depicts departments and reporting relationships. The design of a health care institution's quality management function should address ways that responsibility for quality can become part of every person's role and ways that various committees and communications can foster better coordination in the health care organizations of today, as well as those that will be preeminent in the next millennium.

HISTORICAL BACKGROUND

Quality management, as a named and identifiable function, is a relatively new addition to the activities of today's health care institutions. It evolved from separate areas, including quality assurance, utilization review, risk management, and quality improvement. These areas generally originated as responses of health care institutions to pressures from the external environment.[3]

The first suggestion for an organizational structure to address quality of care in the United States came from Dr. Ernest Codman in the early 1900s. He proposed that each hospital employ an efficiency committee composed of board members, administrators, and medical staff to conduct long-term follow-up on every patient to determine whether the treatment received had been successful. Reviews of unsuccessful treatments were to be the basis for changes in medical practice and hospital operations.[4] At the time, there was a great deal of resistance to Codman's ideas.[5] Nevertheless, the American College of Surgeons incorporated some of his principles when it was founded in 1913. The College subsequently outlined expectations for providing and documenting care in the 1924 Minimum Standard, which formed the basis for its voluntary accreditation process, the Hospital Standardization Program.[6]

While recommendations for different types of quality care reviews by the medical staff appear in textbooks on hospital organization in the first half of the twentieth century, there were no details on the types of administrative support needed. When addressed at all, quality emerged as the responsibility of the clinicians with an emphasis on discipline-specific peer review of the individual provider's actions. For example, MacEachern identified audits of mortality and morbidity as functions of an organized medical staff, but did not include a quality-related department in the recommended organizational chart for a hospital.[7]

In 1951, the American College of Surgeons, the American College of Physicians, the American Hospital Association, the American Medical Association, and the Canadian Medical Association cooperated to form the Joint Commission on Accreditation of Hospitals.[8] During the early years of Joint Commission activity, hospitals were apparently able to achieve accreditation without adding additional hospital staff. As recently as the early 1960s, hospital management texts still did not recommend a quality-related department as part of hospitals' organizational structures.

The 1972 Professional Standards Review Organization (PSRO) legislation contained in Public Law 92-603 introduced a new emphasis: utilization review. The legislation delegated peer reviews of quality and utilization to hospitals, with an emphasis on containing costs rather than improving quality.[9] With the formalization of quality and utilization review from the federal mandates, textbooks on hospital management began to include these functions, often as the responsibilities of separate departments.[10] It became apparent by the late 1970s that the PSROs were ineffective in controlling costs or quality,[11] and they were phased out through a series of legislative acts in the 1980s. The Utilization and Quality Control Peer Review Organizations, more commonly referred to as Peer Review Organizations (PROs), succeeded them.

Despite a lack of evidence that externally imposed utilization review programs eliminate unnecessary health care costs, utilization review increased in the 1980s. The incentives contained in Medicare's prospective payment systems and, stimulated by the demands of business to slow the spiral of health care costs, the requirements of other reimbursers and various state coalitions were responsible for this increase. To accommodate the greater demands associated with additional reviews, hospitals generally involved more staff with utilization review.

Hospital risk management, separate from an insurance-related function, emerged during the mid-1960s in response to an increase in litigation. According to Kraus, the 1965 landmark case of *Darling v. Charleston Community Memorial Hospital*, in which a hospital was held negligent for acts of its medical staff, precipitated health care's first medical malpractice crisis.[12] By the early 1980s, loss prevention and claims management had become a stable hospital activity, usually accomplished through a separate "risk manager."[13] A variety of case-finding methods (e.g., generic screens, occurrence screens, flags, sentinel events) evolved during this period, spurring the use of databases maintained through personal computers with programs developed in-house, or through purchased proprietary systems.

Kraus suggested that the second malpractice crisis occurred in the middle of the 1980s.[14] It was characterized by escalating jury settlements, increased malpractice premiums for physicians, the abandonment of practices, and the withdrawal of insurance companies from the medical malpractice market. In terms of staffing for risk management, it became more common for hospitals to hire lawyers as employees.

In the late 1980s, several health care experts introduced the continuous quality improvement philosophy and approaches to health services organizations,[15-17] stimulating considerable debate and reflection about quality. O'Leary suggested that the original choice of the word "assurance" was semantically unfortunate, since quality could not be assured, but only improved.[18]

The adoption of continuous quality improvement programs in hospitals brought additional changes in organizational relationships and structures. A continuous quality improvement oversight committee that was separate from the existing committees frequently guided the transformation, providing a forum to support the activities occurring in teams and departments. The quality improvement staff and initiative often operated parallel to traditional activities during the early stages of implementation.

In hospitals where the quality improvement initiative has matured, the newest organizational structures encompass both traditional and continuous approaches to quality. For example, at LDS Hospital in Salt Lake City, Utah, the quality management department includes the areas of quality assurance, utilization review, continuous quality improvement, and risk management.[19] At the University of Michigan Hospitals in Ann Arbor, Michigan, the same organizational leaders recently began to oversee the total quality management and quality assurance activities.[20]

Health care facilities and other players in the industry with large databases (e.g., billing data or abstracted medical records) launched a variety of efforts in the 1980s. Case-mix and severity adjustment approaches received substantial attention as components of systems for identifying cases for further review and profiling the care given by individual providers. Despite the lack of independent evaluations or standards upon which to evaluate them,[21] use of these systems rapidly expanded. Such systems were often implemented without practitioner involvement and receipt of the data for review and comment.[22] Organizationally, the increased importance of medical records and information systems resulted in more frequent interactions between the departments responsible for these functions, as well as with the quality, utilization, and risk management services. In a few health care institutions, some or all of these medical information–quality management functions were combined into one department, or their staff supervisors reported to the same senior administrator.

In some institutions, the emergence of the paid medical staff manager role separate from the traditional, elected medical staff leadership positions has affected the organizational structure and activities of the quality management function. While the position may be part-time or full-time and its level in the organization may vary (e.g., vice-president, associate vice-president, or director), the person in the position usually provides the administration with a medical staff perspective in strategic and operations management, and establishes an additional communications link between the administration and the medical staff. Because quality, utilization, and risk management issues are often of direct concern to the medical staff and the administration, medical staff managers spend a considerable amount of their time with the quality management professionals. As a result, some institutions have changed their organizational design so that the quality management department staff report to the medical staff manager. Although this probably decreases the organizational visibility of the nonphysician quality manager, the design may be appropriate for organizations in which the medical staff have lagged behind the rest of the institution in embracing contemporary quality management approaches.

Another emerging change in health care institutions' quality management structures concerns the governing board. In recent years, the use of a separate board committee for quality has been recommended.[23,24] Like the board finance committee, this group should be able to provide leadership in the quality management planning for the organization and should develop greater sophistication in the oversight of quality than is generally possible with the full board.

Finally, the use of discipline-specific and site-specific processes in quality management has influenced organizational structures. With few exceptions, each disciplinary department (e.g., nursing, physical therapy, radiology) has conducted assessments and interventions related to the quality of care provided by its members. Summaries were then forwarded to the institution's interdisciplinary committee. As continuous improvement approaches are adopted, more systems-related issues are being addressed in cross-functional teams. Historically, quality has also been reviewed by site, for example, by hospitals, ambulatory care facilities, occupational health services, home care agencies, and long-term care settings. While reimbursers and insurers have received data from the various sites of service, they have not usually analyzed the care of individual patients over multiple years from different types of providers, perhaps a step we will see in the future.

Table 5–1 summarizes three important phases in health care quality management. In the first phase, the structures and practices of quality management were externally driven. As new health care reforms are introduced, it is reasonable to anticipate that quality management programs and structures will continue to change. Preeminent health care organizations, however, are already beginning to rely on internal assessments and to respond to community needs in advance of any external mandates. In the current, second phase, the era of parallel activities and structures is ending; there will be more integration of staff in the areas of quality management and information, as well as across disciplines. One of the challenges for the future will be to balance the resources devoted to discipline-specific and cross-functional approaches to quality management. The next millennium—the third, speculative phase—will be characterized by community-based, longitudinal, patient-focused approaches to assessing and improving quality with the involvement of providers from the various care settings.

Table 5–1 Health Care Quality Management Phases

	1970s and 1980s	1990s	Next Millennium
Driving force	External regulation and accreditation	Many organizations still responding to external influences, but preeminent health services organizations beginning to evaluate internal operations and community needs	Community needs
Quality management and information areas	Separate	Some integration	Highly integrated with a focus on the community level, longitudinal analyses of quality of care and services
Professional activities	Parallel, discipline-specific	Some integration, but still within discipline-specific hierarchical reporting structures	Fully integrated work teams, flatter organizational structures that are client service-based rather than discipline-specific
Settings	Separate, though some consolidations into vertically and horizontally integrated systems	Fewer separate organizations, though alliances generally driven by present or anticipated changes in financing mechanisms	Highly linked services with a community focus

INTERNAL ORGANIZATIONAL ASSESSMENT

The first step in linking the design of the quality management program with overall effective and efficient organizational performance is an assessment of the organization. The quality improvement framework underpinning the assessment process is depicted in Figure 5–1.[25] The components of the figure (i.e., culture, information/measurement, and actions for improvement) are the commonly identified elements from the quality improvement literature. The metaphor of the three-legged stool is an attempt to capture an often overlooked point of organizational development: balance. Indeed, in organizations that have experienced difficulties with quality improvement, the root of the problem can usually be traced to

1. lack of an assessment of each of these three components before initiating a broad development process
2. overdevelopment of only one or two of the components

In some organizations, for example, there are many individuals who have developed skills in gathering and displaying data. Because of a lack of attention to actions for improvement, essentially problem-solving techniques (e.g., delivering clear, unambiguous communications; listening to understand; resolving issues with the people involved rather than through destructive dyadic conversations with third parties), however, these individuals have not made much progress in addressing long-standing, systemic problems. The issues addressed by the cross-functional teams tend to be superficial and have little impact as the members "play it safe" and avoid challenging the status quo. In such cases, the *reality* of the organization's culture and the problem-solving abilities of the members interact to maintain pathological patterns despite the public *rhetoric* about quality improvement.

Some problems with quality improvement processes stem from the lack of a prior, objective organizational assessment in combination with the overdevelopment of statements concerning culture. This has been particularly troubling when employees were first attracted to an organization because of the values for which it was recognized in the community. Both those employees and community members may misunderstand the replacement of traditional terms with contemporary terms. Meanwhile, the wordsmithing obscures the need to address inconsistencies in behavior stemming from a lack of problem-solving skills to complement either the traditional or the new cultural statements.

Given the current confusion with terms, it is necessary at this point to clarify the terms *quality management* and *quality improvement*. In this discus-

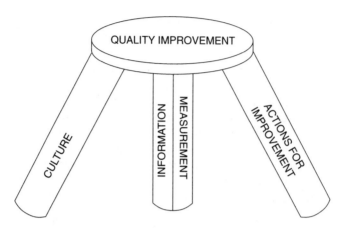

Figure 5–1 Components of Quality Improvement. *Source:* Reprinted from *Improving Quality: A Guide to Effective Programs* by C.G. Meisenheimer, p. 105, Aspen Publishers, Inc., © 1992.

---

sion, the term *quality management* encompasses quality improvement, including the three elements of culture, information/measurement, and actions for improvement (problem solving), and traditional approaches to quality assessment and intervention. Although it is possible to improve an institution's culture and problem-solving capabilities, as well as to learn new quantitative approaches without ever using the word *quality*, quality management and quality improvement will continue to be framing concepts of this discussion. In summary, successful organizations will first *assess* and then proceed to *develop* their organizations to achieve *balance* among the three quality improvement components. Over time, they will achieve new levels of sophistication, using language and visual metaphors appropriate to the setting.

Assessing Culture

Although frequently used in discussions about quality improvement, the term *culture* is often not defined. In this assessment framework, two questions about culture are of interest. First, is the organization committed, by day-to-day actions, to developing the quality culture? In other words, is there more than rhetoric? Second, is the organization making progress in developing and maintaining covenantal relationships? According to DePree, "covenantal relationships fill deep needs, enable work to have meaning and to be fulfilling. They make possible relationships that can manage conflict and change."[26] Organizations characterized by an absence of covenantal relationships may encourage and ignore abuse of persons in staff interactions.[27]

In health care institutions, assessments of organizational culture are usually limited to items contained in periodic satisfaction surveys of patients, employees, and physicians. In this discussion, the cultural components of the content and context of the survey are of interest. After briefly discussing examples of problems some organizations have encountered with such surveys and the indicators of culture which they suggest, several additional strategies for gathering information on culture are suggested.

Used with careful consideration of the (1) purpose and possible actions, (2) development of questionnaire items, (3) analyses of response rates, and (4) feedback and follow-up mechanisms, satisfaction surveys can be a valuable source of information. However, each of these areas can present pitfalls, which may provide insight into the implicit beliefs and culture of the organization (Table 5-2).

The first problem encountered with satisfaction surveys arises from unclear purposes. As a glass that is half-full, for example, an internally conducted monthly or quarterly satisfaction survey of all or a selected number of admitted patients is a sincere effort to gather useful information. As a glass that is half-empty, it is often undertaken because "everyone else is doing them," the governing body or external review agencies expect it, or management wants to create an image that may or may not translate into concrete actions to improve systems. Organizations that consistently have less than a 25 to 30 percent response rate to such questionnaires should seriously reconsider their use. To continue with them under such circumstances will only result in frustration, waste scarce resources, and suggest that the surveys are being done for questionable reasons.

Individual departments often have very important clinical reasons for contacting patients; for example, the nursing staff of an ambulatory surgical center may telephone the previous day's surgery patients to check on signs of complications and reinforce discharge instructions. Often, however, departments are seeking direct feedback by telephone or mail as a defensive response to the vagaries and low response rates of the institutional patient satisfaction survey. As a result, "customers" become inundated with multiple contacts from the health care institution. Like patients, physicians are increasingly inundated with surveys. In organizations where multiple departments want to contact "customers," scheduling departmental contacts on a rotating basis can be a viable compromise.

Table 5–2 Satisfaction Survey Pitfalls and Pragmatic Approaches

Issue	Manifestations of a Pitfall	Pragmatic Approach
Purpose	Performed by other organizations and/or expected by external agencies; most likely to be the case when the survey is continued despite consistent response rates of less than 25–30%.	Reconsideration of survey style (e.g., less frequent surveys with multistage follow-up with nonrespondents).
Number of contacts	"Customers" receiving more than two contacts from various parts of the organization; most likely when individual departments are concerned about survey conducted by the organization.	Annual plan to allow individual departments to contact clients on a rotational basis, and an annual or biannual organizational survey.
Resources	No prospective plan for providing resources to address issues identified in survey.	Targets set for degree of satisfaction below which resources will be available for improvements.
Participation	Departments collect information on other departments.	Opportunity for all involved parties to participate in survey design and receive formal feedback.
Information	Asked to rank-order departments, communicating competition.	Focus on ratings, not rankings.
	No demographic information, making it difficult to assess issues fully and develop targeted interventions.	Demographic information obtained and sensitively used to improve the organization.
Distribution	Haphazard.	Targeted.
Attention to the interests of the target group(s)	Coarse-grained, shotgun approach of using one survey for several different groups.	Fine-grained, laserlike approach of surveys modified for the needs of different groups.
Item nonresponsiveness	Not assessed.	Important part of the analysis for identifying sensitive areas within levels of the organization.
Feedback sessions	Presentation of detailed, department-specific information in a large-group forum, promoting competitive atmosphere.	Presentation of aggregate data in large group, with organizational and area comparisons for individual review between manager and administrator.

Related to the selection of an appropriate purpose is the issue of a plan for actions that may be necessary, depending on the results of the customer surveys and the resources available. Dollars should be set aside to address the required improvements. For example, if physicians are asked what types of assistance they need to provide care for additional Medicare and Medicaid patients, the administration needs sufficient resources to implement some of the recommended actions. It may be helpful to preset numerical values for items in surveys, values that will trigger an investment of resources for improvement (e.g., less than 75 percent of the respondents expressing at least some satisfaction with a particular service). An organization's failure to commit resources for improvements in response to customer feedback may be another indication that the organization is undertaking the survey task for the sake of image rather than substance.

Questionnaire items and their development are often a source of insight into an organization's values. For example, questions on an employee survey that ask respondents to rank-order the responsiveness of other departments suggest that competition, rather than collaboration, is a part of the culture. Surveys that are initiated within one discipline, but include questions about other disciplines also provide insight—particularly if members of the other disciplines (1) did not know about the survey and, consequently, did not have an opportunity to contribute to the construction of the items about their services, and (2) hear criticisms about their services only after the release of these criticisms to members of the original discipline. Such occurrences suggest a lack of openness and limited problem-solving abilities among the organization's members.

Questionnaire items must include sufficient demographic information to ensure that the information collected is meaningful. For example, one institution asked a question on an employee survey about the adequacy of orientation. No information was requested about the date on which the employee had been hired (e.g., year). The education service had recently been reconfigured, and the orientation had been substantially revamped. On the survey, many employees reported a lack of satisfaction with orientation. Because an employee only goes through orientation once, whether 30 years ago, 10 years ago, or yesterday, there was little that the new department could do in response to the survey's results. It would have been more meaningful in such a situation to ask new employees whether they were satisfied with the program at the end of orientation and to ask whether the program had prepared them well at periodic intervals after the completion of orientation. For the purpose of improving the orientation process, systematic information could be gathered from additional sources (e.g., managers and peers) on the adequacy of the orientation of new employees during the previous 6 months to 1 year.

The distribution of questionnaires sometimes reveals the culture's inclusiveness. In one hospital, an in-house contingent personnel pool provided a high proportion of the registered nurse staffing. When a nursing satisfaction survey was conducted, the contingent nurses were to be included. The first-line nursing managers generally made an effort to give the survey to each nurse on the regular unit staff, but simply left the surveys for the contingent staff in the central area where they checked in for their shifts. Furthermore, the first set of questions concerned "the" unit manager, and contingent staff related to many of the managers. Therefore, both the method of delivery and the initial survey content indicated to the contingent staff that there was not a sincere effort to include them; because few responded to the survey, there was a loss of valuable information for the organization.

To elicit responses from diverse groups, the first step is to design a questionnaire with a subset of the items relevant to each group, such as the contingent staff and unit staff in the example given earlier. This is an easy task with today's word-processing programs. Second, it is wise to personalize the delivery, perhaps by mailing the survey to the home addresses of the contingent staff. Although that may sound more expensive, considerable cost went into the process of surveying contingent staff with little usable results—the old "penny wise, but pound foolish" adage. Similar problems have been encountered by hospitals surveying physicians. Many hospitals have a core set of physicians who are the high-volume admitters and a set who are infrequent users. It is reasonable to expect that they have different perceptions of the organization. Rather than the shotgun approach of one survey, a laserlike approach of two surveys conducted at the same time with relevant questions would yield greater insights. Results of the two surveys would be compared to determine if different groups have differing perceptions of the organization's core characteristics.

An often overlooked area, particularly with employee surveys, is the pattern of nonresponse for each item. Some employees may not finish the survey, perhaps due to time constraints. As a result, the response rates for the later questions are consistently lower than those for the earlier questions. In future surveys, it may be necessary to alter the circumstances so that there is sufficient time for completion. Of greater concern from a cultural perspective is selective nonresponsiveness, which indicates that employees perceive certain topics as risky to answer. The leaders, perhaps with an outside consultant, need to examine the items with lower response rates to determine if there are similarities in the concepts that they address, such as communication. In addition, if demographic information (e.g., position, unit, shift) was obtained, the nonresponse items should be analyzed by those categories. When one organization conducted an employee survey that included several levels of management, part of the demographic information included level of position and department. Some of the questions asked for comments about leadership style in the department. This was a fear-filled organization, however, and there was a lack of trust between the first-line and middle managers. In the belief that any negatives would be attributed to them, the first-line managers simply did not answer those questions. Because there was no assessment of the low response items, the executives congratulated themselves on what they perceived as relatively positive results. Meanwhile, the chasm between the levels of management continued to grow.

As noted earlier, any group about which questions are asked should be included in the planning of the survey and should receive the feedback concurrently with other groups. If a survey is done over several departments, such as an employee survey, it may be appropriate to review the aggregate results, general trends, and organizationwide interventions in a large group forum. Then, the manager and administrator can discuss an area's specific results and plan additional actions. It may also be helpful to display the results in the aggregate and to produce for each department a report of the department and aggregate results, perhaps with means and standard deviations or some other control chart displays. Sharing the feedback for the first time in a large forum with each department's results displayed creates a competitive, rather than a constructive, atmosphere. All departments, even the best, usually have opportunities for improvement as a result of surveys of patients, physicians, or employees, however.

Besides examining the content and context of surveys, it is possible to gain insight into the organizational climate by analyzing the mission, vision, and value statements to determine if they are incorporated into the daily life of the organization. Administrative actions should be consistent with the cultural rhetoric. One organization had initiated a quality improvement process that included rewriting the mission and other statements. Shortly thereafter, the first-line management role was updated and expanded. The job description was developed with great fanfare, including discussions with "stakeholders" to develop shared expectations and understandings about the role. The expectations for quality contained in the document were limited to those required by accreditors and regulators, however. Consequently, the managers neither allocated much personal effort to quality improvement activities nor developed the needed skills among staff. The organization missed an opportunity to be consistent with the change to a continuous improvement philosophy.

Calendars and appointment books can reveal a consistent commitment to quality. What proportion of time is devoted to quality at different levels of the organization? Of that time, how is it distributed among quality planning, quality control, and quality improvement?[28] Senior executives would be likely to spend a high proportion of their time involved with quality planning. Quality improvement would command a large proportion of the time of middle and first-line managers. All staff would be involved with quality control, with the proportion perhaps determined from work sampling and job descriptions rather than calendars.

A qualitative review of meeting attendance, cancellation patterns, start times, and minutes may also provide insights about the organization's commitment to quality. At the organizationwide quality committee, are some departments' reports always deferred at least once due to lack of preparation? Are senior executives routinely present at the meetings where they are members? At the governing board meetings, is the quality report the last item on the agenda, with attachments referred to as people are gathering their belongings and beginning to depart?

Clearly, there are many ways to assess an organization's commitment to quality. It may be most helpful to conduct a multifaceted review on an annual basis as part of routine strategic management activities. The first task may be a brainstorming session to develop additional measures of the organizational commitment to quality. In an organization truly committed to improvement, such a process can provide new insights for organizational development.

Assessing Actions for Improvement (Problem Solving)

Organizational leaders use information to solve problems and develop their organizations. Figure 5–2 summarizes four caricatures of information management styles that are barriers to problem solving. The "Just the Facts" style in Figure 5–2 is one in which only the data are of interest. There are two variations on this stance. In the first, the leaders want no interpretation of the data, a view they often take in the name of "professional autonomy." Clinical disciplines are particularly known for this approach. The second variation on this theme is an often untested belief that professionals who have had extensive graduate level preparation will know what the data mean and will do "the right thing" if they have "all the facts."

At the opposite end of this continuum are the "Don't Bother Me with the Details" types, most often found among senior administrators. They have an aversion to personally reviewing data and are always asking for "the

Designing Quality Management Programs 69

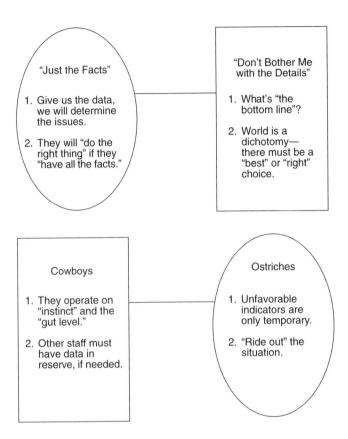

Figure 5–2 Caricatures of Problem-Solving and Information Management Types

bottom line." They see each issue, and the world in general, as a dichotomy: one of the two choices must be "right" or "best." Their view of the world does not take into consideration nuances about decisions, diversity in perspectives, or a range of possible choices.

On the next continuum, the "Cowboys" share the aversion to data of their "Don't Bother Me with the Details" colleagues. Their view of the world is even more extreme, however. They want to make quick decisions on "instinct," at the "gut level." From their perspective, only indecisive folks use data, but, of course, someone must be ready to trot out data when external agencies visit. Finally, the "Ostriches" consider data and trends, but fundamentally regard unfavorable indications as temporary aberrations to "ride out" because "things will get better."

Given that organizations do have some of these tendencies, there are several practical ways to assess an organization's problem-solving capability. One general approach is to combine a review of the formal quality reporting paper trail with interviews. As with the assessment of the organization's culture, this analysis can be part of the annual strategic management review. The process described in the following may be helpful at both the organization and department levels.

The content analysis of the quality reporting documents requires the reviewers to consider only the information in the documents, not what they "know" occurred. For that reason, it may be best if individuals who are not routinely involved in the quality management reporting process conduct this review. The fundamental question to be answered in this review is, Does the organization follow through on issues in a timely manner, demonstrating the espoused values?

The first step is to identify five to eight issues to track through the improvement process. This is probably best done by reviewing the minutes of the quality meetings from the operational areas (e.g., the unit, department, or service levels). The issues should represent a variety of individual areas and interdisciplinary concerns. The specific steps for the content analysis of quality reporting documents are as follows:

1. Diagram the groups and individuals to whom the issues were referred, noting the amount of time that elapsed at each step.
2. Identify the type of data and analyses used in generating solutions.
3. Classify the types of solutions employed.
4. Determine whether and what type of follow-up was implemented to prevent recurrence of the issues.
5. Determine if the issues recurred. For recurrences, note how subsequent interventions differed from the first interventions.
6. Throughout the review of documents, use a check-sheet to note the frequency with which the espoused values are mentioned.

If there are management meetings separate from quality meetings at the first-line and middle management levels of the organization, those minutes may also be of interest to identify (1) the proportion of time that quality occupies in the managerial meetings and (2) the types of issues that are classified as "management" without a quality component. This review may highlight whether the techniques of quality improvement are broadly or narrowly applied.

When done in an open spirit of inquiry, document content analysis can provide several insights, promoting further learning and organizational development. Management is usually surprised at the amount of time consumed by the issues, but should not react by encouraging speed for the sake of speed (e.g., mandating more meetings, setting time limits on issues). Rather, management should address questions such as the following:

- Are there patterns of communication that could be improved?
- Do certain committees or individuals tend to delay resolution?
- Are there recurring problems, sometimes in different areas of the organization, that perhaps require more focused attention by a broadly representative intervention team?
- Is there periodic, proactive follow-up on previously identified issues rather than a reliance on crisis management?

Organizations commonly take several actions to address the time lapses and spur follow-up. Specifically, the forms for documenting quality management issues and meeting minutes are usually changed to include multiple time cues. On the reporting forms, headers specify the time of issue identification, planned and actual initial data gathering and evaluation, planned and actual dates for interventions, planned and actual dates for evaluation. In minutes, the header may be modified to include the issues that have been referred to other committees or individuals and the date of referral. The referred items remain in the header under "Old Business" or "Referral" until there is a resolution.

As a result of the quality document content analysis, senior administration may recognize that the quality management process requires their active participation. Often, the vice-presidents, chief operating officer, and chief executive officer are unaware that issues are cycling through various levels of the organization. As the reporting formats become more specific, organizations may find it helpful for senior managers to see the reports. Alternatively, particularly in large organizations, it may be decided that issues should be taken to a senior administrator after a specified amount of time has elapsed without a response from another part of the organization.

Organizations that are trying to "empower" employees are often surprised to learn that employees frequently refer action steps to individual managers or management forums for decisions. Or, a pattern may emerge in which higher levels of management are not supporting solutions that require resources. In other cases, middle and first-line managers may be inappropriately abdicating management responsibility to committees. For all of these situations, it is necessary to develop deeper shared understandings of delegation and participation. In addition, there is a need for clear resource parameters within which committees can work.

Other interesting findings may include (1) the infrequent attention to values in resolving issues, (2) the use of only a few approaches to gathering and displaying data, and (3) an overuse of some interventions. It is not unusual to find organizations in which the first intervention is always group education, for example. The education sessions are often poorly attended, and the problem recurs. The "training trap" does not address the underlying systemic issues, nor does it take into consideration that the individuals most in need of such programs are usually the ones who do not attend.

In some organizations, employees invest considerable time in gathering data to solve issues with seemingly obvious solutions because of a perception that a certain individual or department wants that type of data-based approach. As an old adage suggests, common sense may not be very common. For such situations, however, some reality testing is clearly required.

Assessing and Planning Information/Measurement

The sequence of the words in "information/measurement" reflects a philosophical bias: the organization needs to begin with a clear picture of the desired information end products. Fundamentally, there are two different levels of information needed in health care settings. The first level concerns the oversight of aspects of the institution's quality selected as critical indicators of organizational performance. Clinical and service quality indicators at the organizational level may include readmission rates, average length of stay, and employee turnover.

The second level of required information is at the operational level (e.g., department, unit, or service). At the department or unit level, routine reports concern overtime and supplies, since these aspects of care are most directly under the area's responsibility. Quality information at this level includes evaluation of the adequacy of patients' discharge teaching (which may affect the readmission rate); daily progress on critical pathways, as well as deviations from standard time frames for support and consultation services (which, in time, may affect the length of stay); and adequacy of staffing based on patient need and volume (which may eventually affect turnover). Departmental information can also be aggregated into organizational level measures. Juran recommended that the technical departmental reports be transformed into summarized reports, often using ratios, for use at higher levels of the organization.[29]

To determine the end product with respect to information at the organizational and operational levels, a series of simple questions can be answered:

1. What types of information are needed for oversight at the organizational level?
2. What types of information are needed for process improvements, generally at the operational levels?
3. When is the information needed? For example, adverse events need to be reviewed immediately, while information for routine monitoring and evaluation within a specific department may be monthly; administrative and governance levels may need summarized reports only on a quarterly basis.
4. Who needs the information? The level responsible for action should receive the information first.
5. In what format? Research suggests that people engage in more problem-solving activities with tabular formats, but that identification of trends and recall are facilitated by graphic reports.[30,31]

Figure 5–3 provides additional detail for answering several of these questions and also indicates when data collection (the "measurement" in information/measurement) should occur. When immediate action is necessary, data gathering is concurrent. In hospital or nursing home settings, for example, specific individuals such as quality/utilization review staff who make unit rounds may collect information and give it to the appropriate individuals responsible for action in either a centralized department (e.g., risk management, patient services, personnel) or decentralized area (e.g., the nursing unit, pharmacy, surgery). If immediate action is not necessary, it is generally acceptable to gather the information retrospectively and to produce aggregated trend reports. For specific data that do not require professional judgment, centralized abstracting by means of current and emerging medical records abstracting systems and financial/claims databases may be an efficient approach. If professional peer review judgments are needed, individual case review, most likely conducted within the decentralized operational areas, is appropriate. All data obtained should be available in a database for later queries and production of more summarized reports.

If an organization elects to have most of the data gathered through a centralized, retrospective abstracting process, there are three general approaches:

1. The operational areas can query the database directly and generate the data that they need for analysis and reporting.
2. The data desired by the operating units can be extracted centrally, perhaps by the quality management department; analysis can then

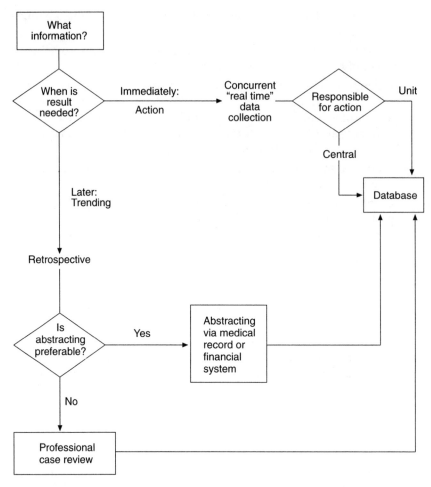

Figure 5–3 Information/Measurement

---

be conducted and summary reports produced by the operational areas.
3. The data can be extracted and a preliminary analysis conducted centrally by the quality management department or, in institutions with sophisticated management information systems, via predetermined, perhaps automatic, algorithms. When there are indications of an adverse trend, the information is brought to the attention of the leaders of the operational area.

There are clear philosophical and resource differences between these approaches. The first two approaches place the responsibility for analysis and reporting with the operational areas. Furthermore, the first approach requires that the database be accessible to the operating areas and that at least some staff members in each operational area have the necessary computer skills and time available to work with the database. If this is not practical, the operational areas need a central department to provide data support. The third general strategy strives for some economies of scale between the central and decentralized quality management activities. A mechanical system or central department carries out the ongoing monitoring, which allows the operating units to concentrate on improvements rather than the generation and review of routine reports.

It may be valuable to address the three information/measurement approaches in the annual strategic management review. The institution needs an overall philosophy concerning roles and responsibilities for data gathering, analysis, and reporting. The general philosophy should then guide the application of resources for the development of information systems and the types of skills needed by centralized and decentralized staffs.

In summary, a variety of diagnostic tools have been discussed for assessing the quality culture, capabilities for taking action, and approaches to information/measurement. These assessments can be conducted during a periodic strategic management review. It also may be practical to couple these assessments with the annual review of the quality management program and plans. Assessments are only a starting point. In the next and last major section of this chapter, additional suggestions are offered for designing effective and efficient quality management programs for the future.

INTERNAL ORGANIZATIONAL EVOLUTION

With the assessments complete, an organization is ready to plan for the future. One transcendent issue to be addressed in planning the quality management organizational structures and processes is the paradox between the need to make quality everyone's job and the need to make the quality function distinct and highly visible to internal and external constituents. After discussing this paradox and ways to resolve it, three structure and process issues that need to be considered in designing the organization's future quality management function are identified.

Having quality belong to everyone requires concentrated attention to aligning job descriptions and evaluations, as well as opportunities for participation in data gathering, analysis, and intervention. As they began quality improvement initiatives, health care settings often ignored the linkages with personnel processes. Quality management was not part of the educational experiences of most senior executives or members of governance, however, and considerable development must occur at these levels if quality is to be consistently incorporated into fiscal planning and strategic management.

Because of the complexities of quality management, some health care organizations have created a senior executive quality role in recent years. The chief quality officer role is analogous to that of the chief financial officer. The potential advantages of such a role include having a "resident expert" who can stay abreast of changes in quality management, raise issues related to quality in the making of internal decisions, and represent the institution in external forums. Reliance on the designated chief quality officer may create a situation in which other senior executives gain only a superficial understanding of quality management, however. This is particularly troubling when chief executive officers and chief operating officers essentially abdicate their roles in leading the quality improvement process, leaving all responsibility to the individual in the chief quality officer role.

The first step to resolving a paradox is to recognize that it exists. One solution is to commit to both approaches: emphasizing that quality is everyone's job and employing a chief quality officer. In this two-pronged approach, the chief quality officer should have a key role in determining how quality is incorporated into the job descriptions, training, and evaluation of staff at all levels of the organization.

Another way to resolve a paradox is to use different strategies at different points in time. To develop the organization's culture, information/measurement capabilities, and problem-solving skills, and to link them with the personnel processes of the institution, it may be useful to establish the chief quality officer role as a time-limited, transitional position (e.g., for 5 to 10 years). Of course, an individual who accepts such a transitional role needs to consider whether the position fits with his or her long-term career development goals. Furthermore, the institution should actively seek other administrative positions that may be appropriate for the individual who completes such an assignment.

The presence or absence of a chief quality officer is the first of three organizational structure and process issues that must be considered in designing the quality management function: reporting relationships. The chief quality officer role may be either a staff or line position. In a staff position, the chief quality officer may find it easier to serve as coach and cheerleader for the quality improvement process. In addition, it may be easier to make the role time-limited if no other quality management personnel report to the incumbent. If the chief quality officer role is a line position, staff from the quality, utilization review, and risk management areas, and, perhaps, the nontraditional areas of medical records and information systems in organizations pursuing an information-intense strategy, may report to the incumbent to achieve economies of scale and scope in designing the quality management function.

In an institution that does not have a chief quality officer, the argument for achieving economies of scale and scope in activities such as data collection suggests keeping at least the traditional quality management areas together rather than fragmented, for example, having quality the responsibility of a vice-president for medical affairs or patient care services; utilization review, that of the chief financial officer; and risk management, that of in-house legal counsel. Since form should follow function, the reporting relationships must be appropriate to the history and culture of the institution. Of course, the assignment of the quality management function (e.g., finance, clinical services, and in religiously sponsored institutions, mission effectiveness) communicates a message about the institution's values and priorities to internal and external customers.

The second structure and process issue to be considered in designing the quality management function concerns the roles and responsibilities of the various settings (e.g., hospitals, physician's offices, home care, nursing homes) in the joint management of clinical and service quality for individuals and groups. Regardless of the specific health care reforms that may be introduced at state and federal levels, there is a substantial and sustainable movement in the direction of mutual dependency between care settings. Whether through delivery systems, financing mechanisms, or a combination of the two, the client will have a more seamless care system. The next-generation information systems, which can track care longitudinally by using a single patient identifier, will support this evolution. Relaxation of the barriers to collaboration among health care providers from antitrust regulations, already accomplished in several states, will also stimulate more patient-focused initiatives. In addition to quality management reviews *internal* to each setting, forums for review *across* the settings will emerge. To manage these collaborative quality management programs, it is possible that the chief quality officer role will evolve beyond individual institutions.

Information for quality reporting may flow through five levels, from individual performance feedback through governance level (Table 5–3). Each level has a different purpose in the organization and needs quality

Table 5–3 Quality Management Reporting

Level	Frequency	Content	Format
Individual	As often as possible; monthly basis highly desirable, minimum of annually	Critical items participatively selected, encoded for confidentiality	Tabular, with comparative data for department, organization, previous period, same period previous year
Interdisciplinary unit/team	Monthly	Per quality activity with an individual who "owns" the data gathering and preliminary analysis leading the discussion	Tabular, with comparative data for organization, previous period, same period previous year; occasional focusing graphics
Executive leadership	Monthly	Sets of ongoing and annual critical success factors Emphasis on (1) cheering and changing and (2) prospective planning for quality with new and discontinued services	Focusing graphics
Interorganizational community	Monthly	Review of a set of critical success factors emphasizing longitudinal analyses of quality of care and services	Summary tables, focusing graphics
Governance	As often as board or board quality committee meets	Sets of ongoing and annual critical success factors Emphasis on (1) cheering and reviewing persistent unresolved issues and (2) prospective planning for quality with new and discontinued services	Focusing graphics

reports of an appropriate frequency, content, and format. Two general principles underlie this approach:

1. Greater frequency of reporting and detailed, tabular displays are appropriate at the individual and operational levels.
2. More summarized and exception report formats with graphic displays are appropriate at the administrative and governance levels.

Individuals who are a potential source of variation should receive performance feedback for quality activities. There has been considerable confusion in the health care field on this point. Although most obstructions to quality improvement are systems issues and improving average performance results in greater gains than does looking for "bad apples" to punish, the fact remains that individual performance matters. As individuals, most people appreciate feedback and want to improve, provided that the institution's culture is based on covenantal relationships. Being able to work with specific individuals for remediation and improvement also eliminates the "training trap" of education for everyone as a quality management intervention. Not every quality activity can or should be tracked at the individual level, but there should be an iterative, participatory process for selecting a manageable number of quality indicators where individual identity will be gathered and personalized feedback given as often as possible.

In the feedback process with individuals, it is valuable to provide aggregated data that permit comparisons to others in the same personnel category from the same unit or department and, if possible, from the total organization and external sources. For the sake of confidentiality, individual identities should be encoded in a numerical system that is changed at least annually. Detailed, tabular formats are probably the most useful to the individual. Comparative data from earlier periods, both the previous period and same period the previous year, are also desirable. Comparisons to the same period in the previous year allow consideration of the influence of seasonal trends, an issue frequently neglected in health care quality reporting.

Ongoing or time-limited unit improvement teams need to meet at least monthly to maintain momentum. Members need to have data reports far enough in advance to prepare for the session in order to use the meeting time constructively. The meeting should focus on interpretation of the data, discussion of any necessary actions steps, and evaluations of previously taken actions. To facilitate this process, one or two individuals should "own" specific indicators and should be prepared to lead the related discussion. If these principles are followed, meetings need not be longer than 1 hour (1½ hours maximum). If meetings routinely take longer, one or more of the following problems are probably occurring:

- The leader or members are not prepared.
- Productive meeting management principles are not followed.
- Inappropriate activities are carried out in the meeting (e.g., gathering data, tabulating data). Activities are most likely to be inappropriate when there are marathon 4- or 8-hour quality meetings. By the time data gathering and analysis are completed, the group members are too fatigued to conduct the most important part of the meeting: the professional evaluation.

The unit/team level is the only level at which detailed tables for each quality indicator need to be analyzed. The tables should include the aggregate results for the operational area, the results for the previous period and/or the same period of the previous year, and, when possible, comparative information from the total organization or an external source. Occasionally, the individual(s) responsible for a particular indicator may want to supplement tabular displays with graphics.

Quality reporting usually concerns communicating up the organizational hierarchy, but communicating laterally with the other unit/team members who are not on the committee is also important. Some quality information that provides positive feedback and indicates what actions are under way for areas needing improvements, as well as how to participate in those actions, should be routinely shared. A variety of communication vehicles can be employed, including general staff meetings, newsletters, or postings of minutes and graphics for all to see.

The executive leadership level is interdisciplinary to the degree that individuals at the vice-president level and above have different professional backgrounds. These individuals usually meet at least monthly, and quality reporting should be a routine part of each meeting. At this level, sets of ongoing and annual critical success factors (e.g., employee turnover, specialty and board certifications, grievances, malpractice claims summaries) should be reviewed. Generally, the executives have previously received summaries of progress on indicators from the work groups, and they do not rediscuss them at this level. Rather, the goals at these meetings are to

- check that the organization is "on course"
- cheer successes
- provide resources for "course corrections"
- plan for the future (i.e., develop the quality plans to evaluate new or discontinued services)

The last item requires more explanation. The first part of Juran's trilogy, quality planning, addresses new services.[32] To date, however, health care providers have not taken responsibility for assessing the impact on the community of discontinuing services. As health care becomes more community-focused, there is an ethical imperative to include such assessments under the quality management umbrella.

There is less specificity recommended for the interorganizational community level of quality management reporting, as there needs to be experimentation. In general, this forum includes administrators and clinicians from the service settings most closely involved with quality and probably should meet monthly. The purpose is to provide longitudinal review of patient care and develop smooth transitions for care between different sites. The specific set of critical success factors reviewed by this group may include such issues as satisfaction with services provided "upstream" in the continuum of care (e.g., complete referral information, adequate patient education), review of care provided to clients with chronic conditions, and appropriate screening based on demographic characteristics (e.g., ophthalmological examinations within the last 2 years for all persons with diabetes). Epidemiological approaches offer an important contribution to the population-level view of quality management that should characterize these meetings. The forums should be a vehicle to develop new quantitative and qualitative approaches for assessing quality and improving the average health status of community members.

Governance level quality reporting should take place as often as the full board, and board quality committee in organizations with such a structure, routinely meet. Information should be sufficient to provide oversight without micromanaging. For routine monitoring, the board should become involved when an issue has not been resolved despite numerous types of interventions. Usually, the constraints at such a stage are related to resources. The board should also review the sets of routine and annual critical success factors that were reviewed at the executive level. In addition, information from the interorganizational level is of interest to board members. Pro forma financial and quality plans for new services should be part of the board's review of new projects. Finally, the board should begin asking questions about the impact on the community of discontinuing programs.

CONCLUSION

Churchill reportedly once said, "First we shape our structures and afterwards they shape us." As current quality management designs are evaluated and new ones considered, it is important to remember that structures and processes are only the means to an end, not ends in themselves. In the next millennium, the "end" of community-based care services will be different than the setting-specific, discipline-oriented focus of today. Reaching that "end" will require a variety of strategies for assessing and designing the "means" of the quality management function. Despite the time and energy required to create the new frameworks, the transition to a community focus should offer exciting and energizing opportunities for administrators, clinicians, and quality management professionals.

NOTES

1. S.M. Shortell et al., The Effects of Management Practices on Hospital Efficiency and Quality of Care, in *Organizational Research in Hospitals*, eds. S. Shortell and M. Brown (Chicago: An Inquiry Book, Blue Cross Association, 1976), 91.
2. J. Pfeffer, *Organizational Design* (Arlington Heights, Ill.: AHM Publishing Co., 1978), 25.
3. J. Genovich-Richards, Quality Management Organizational Structures: History and Trends, *Journal for Healthcare Quality* 16 (1994):1.
4. A. Donabedian, The End Results of Health Care: Ernest Codman's Contribution to Quality Assessment and Beyond, *Milbank Memorial Fund Quarterly* 67, no. 2 (1989):233–256.
5. A.G. Mulley, Jr., E.A. Codman and the End Results Idea: A Commentary, *Milbank Memorial Fund Quarterly* 67, no. 2 (1989):257–261.
6. J.S. Roberts et al., A History of the Joint Commission on Accreditation of Hospitals, *JAMA* 258, no. 7 (1987):936–940.
7. M.T. MacEachern, *Hospital Organization and Management* (Chicago: Physician's Record Co., 1935), 84.
8. Roberts et al., A History of the Joint Commission.
9. K.N. Lohr, *Medicare: A Strategy for Quality Assurance*, Vol. 1 (Washington, D.C.: National Academy Press, 1990), 146.
10. See, for example, A. Sheldon, *Organizational Issues in Health Care Management* (New York: Spectrum Publications, 1975), 83, 169–170, 174–175, 181.
11. R.D. Luke et al., *Organization and Change in Health Care Quality Assurance* (Gaithersburg, Md.: Aspen Publishers, 1983), xi.
12. G.P. Kraus, *Health Care Risk Management: Organization and Claims Administration* (Owings Mills, Md.: National Health Publishing, 1986), 2–3.
13. D.J. Slawkowski, Legal Counsel and Risk Management, in *Hospital Quality Assurance: Risk Management and Program Evaluation*, ed. J.J. Pena et al. (Gaithersburg, Md.: Aspen Publishers, 1984), 175–184.
14. Kraus, *Health Care Risk Management*.
15. P.D. Batalden and E.D. Buchanan, Industrial Models of Quality Improvement, in *Providing Quality Care: The Challenge to Clinicians*, ed. N. Goldfield and D.B. Nash (Philadelphia: American College of Physicians, 1989), 133–159.

16. D.M. Berwick, Continuous Improvement as an Ideal in Health Care, *New England Journal of Medicine* 320, no. 1 (1989):33–56.
17. R. Laffel and D. Blumenthal, The Case for Using Industrial Quality Management Science in Health Care Organizations, *JAMA* 262, no. 20 (1989):2869–2873.
18. D.S. O'Leary, CQI—A Step beyond QA, *Joint Commission Perspectives* 10, no. 2 (1990):2–3.
19. Integrating QA and QI: Marriage or Mayhem, *Quality Exchange Newsletter*, Johns Hopkins University (1991):1–3, 9.
20. L.B. Creps et al., Integrating Total Quality Management and Quality Assurance at the University of Michigan Medical Center, *Quality Review Bulletin* 18 (1992):257.
21. L.I. Iezzoni, 'Black Box' Medical Information Systems: A Technology Needing Assessment, *JAMA* 265, no. 22 (1991):3006–3007.
22. M.T. Koska, Physician Practice Guidelines Go under the Microscope, *Hospitals*, February 20, 1990, 32–37.
23. B.S. Bader, Quality Assurance: Shaping up Your Board, *Healthcare Executive* 2, no. 3 (1987):28.
24. A.E. Boehm, Fiduciary Duty and the Quality of Care, *Health Management Forum* 5, no. 3 (1984):61.
25. J. Genovich-Richards, Selecting Topics and Methodologies, in *Improving Quality: A Guide to Effective Programs*, ed. C.G. Meisenheimer (Gaithersburg, Md.: Aspen Publishers, 1992), 104–109.
26. M. DePree, *Leadership Is an Art* (New York: Doubleday, 1989), 51.
27. J. Genovich-Richards, A Poignant Absence: Sexual Harassment in the Health Care Literature, *Medical Care Review* 49, no. 2 (1992):133–159.
28. J.M. Juran, *Juran on Planning for Quality* (New York: The Free Press, 1988), 11–13.
29. Ibid.
30. N.S. Umanath and R.W. Scamell, An Experimental Evaluation of the Impact of Data Display Format on Recall Performance, *Communications of the ACM* 11, no. 5 (1988):562–570.
31. R. Bennett et al., Managerial Rations of Written Compositions: Impact of Information Technology on the Persuasiveness of Communications, *Information & Management* 19, no. 1 (1990):1–6.
32. Juran, *Juran on Planning for Quality*.

» Chapter 6 «

Comparative Performance Measurement for Health Plans

Janet M. Corrigan and Lisa S. Rogers

IN RECENT years, increased emphasis has been placed on the development of a common set of measures to assess the performance of health plans (e.g., health maintenance organizations [HMOs], preferred provider organizations [PPOs]) and the reporting of comparative quality information to purchasers, consumers, and others. Large private employers, whose employees represent a sizable proportion of health plan enrollment, have been particularly vocal in their demands for greater accountability and demonstration of "value," and many small and medium-sized employers have begun to form coalitions to use their joint purchasing power as leverage when contracting with health plans.[1] More and more often, the cost of health care benefits appears to determine the competitiveness of a business in both national and international markets.

Standardized performance reporting requirements are integral components of health care reform proposals and efforts at both the national and state levels. The Clinton Administration's health care reform proposal not only provided for the development of a set of national performance measures, but also required health plans to maintain certain standardized data sets and to report on specified aspects of performance in the areas of access, satisfaction, appropriateness, outcomes, utilization, and financial viability.[2] Although the likelihood of significant national health reform legislation has diminished considerably, similar provisions for performance monitoring and reporting are evident in various state health reform initiatives.[3]

A system of standardized reporting will serve several objectives. Health plans will have access to benchmarking information to identify areas for improvement. The availability of information on health plan performance across a diversified set of performance measures will also provide consumers, purchasers, regulators, and others with the information necessary to select the health plan that best satisfies their needs, thus giving health care plans an incentive to improve their performance and value.

CHARACTERISTICS OF HEALTH PLANS

Falling under the rubric of managed care are many different types of health plans, including PPOs, point of service (POS), independent practice associations (IPAs), and group and staff model health plans. Little, if any, objective data are currently available to assess and compare the performance of these various types of plans. Because of their financing and organizational structure, however, health plans offer the opportunity to assess aspects of performance that are difficult, if not impossible, to assess in less integrated, fee-for-service environments. More specifically, health plans generally serve an enrolled population and provide comprehensive benefits through a defined network of institutional and individual providers.

Enrollment in managed care organizations has grown steadily. In 1993, there were approximately 45 million individuals enrolled in HMOs and more than 121 million in PPOs.[4,5] Having an enrolled population makes it possible to study both users and nonusers of services. For example, instead of counting the number of diabetic individuals referred for an annual eye examination by a physician group practice, it is possible to calculate the proportion of diabetics in a health plan's enrolled population who received an annual eye examination. The second figure is far more useful than is the first in assessing the extent to which a *population* is receiving adequate and appropriate health care.

Population-based measures for geographical areas have revealed wide variations in practice patterns and raised concerns about the appropriateness of care.[6,7] Geographically defined population-based measures are of limited use for ongoing performance monitoring or for consumer/purchaser decision making, however, because there is generally no organizational entity that has accepted responsibility for serving the population and, thus, can be held accountable.

By assessing the performance of health plans with enrolled populations, federal and state governments, as well as community groups, can also gauge progress toward meeting public health priorities. In 1990, the U.S. Public Health Service established national goals for health promotion and disease prevention by the year 2000.[8] For example, because it was estimated that only about 70 to 80 percent of children aged 2 were fully immunized, the following goal was established: "Increase to at least 90%, completion of the basic immunization series among children under age

2."[9] Holding health plans accountable for monitoring childhood immunization rates and demonstrating improvement over time represents a potential strategy for attaining the year 2000 goal.

Because most health plans provide a comprehensive package of benefits, enrollees receive the majority of their care from plan providers (i.e., "in network" hospitals and physicians). Therefore, it is possible for health plans to assemble reasonably complete information on an enrollee's service utilization across various sites and levels of care. Only a small proportion of health plans currently have such information available in an automated system that allows for efficient retrieval, but more are now investing in the information systems necessary to give them this capability in the future.

Enrolled populations that are reasonably stable afford researchers the opportunity to assess changes in health status over extended periods of time. Once again, only a minority of health plans have large, stable populations and the necessary data collection mechanisms to take advantage of this opportunity. As the population ages and chronic diseases increase in prevalence, it is likely that there will be a greater emphasis on the collection and linking of necessary data across various sites and over time to allow for more longitudinal analysis of health care interventions and outcomes.

The health plan has certain advantages over the individual physician practice as the unit of analysis for quality measurement. First, at the level of a health plan, the number of specific types of clinical occurrences is large enough to allow for assessment of numerous clinical issues. In contrast, analysis of individual physician practices must take into account the fluctuating results that stem from small numbers of cases, making interpretation and evaluation of performance difficult.

Second, population- or system-based measures have the advantage of focusing on health plan processes that are not individual provider–based. They recognize that the health care system consists not only of a linked network of physicians, but also of complex processes between and among components of the system. Experience with total quality management in non–health care sectors clearly indicates that the majority of problems are system problems rather than individual operator problems. In health care, improving system processes is as important as improving individual provider performance.

MENU OF PERFORMANCE MEASURES

To provide an adequate picture of health plan performance and to meet the needs of multiple users (e.g., consumers, purchasers, regulators, individual providers, and the health plans themselves), a performance assessment system should satisfy several criteria.[10] First, such a system should include a well balanced set of measures that reflect important aspects of health plan performance:

- accessibility of care
- enrollee satisfaction with care
- technical quality of care
- efficiency and cost-effectiveness
- financial solvency and stability of the health plan

Second, the performance measurement system should include measures applicable to various clinical areas and levels of services, such as preventive care, emergency/critical care, acute care, and chronic care. It should also address major clinical areas, such as pediatrics, obstetrics and gynecology, family medicine, internal medicine, general surgery, mental health, and various subspecialties.

Third, the performance measures should include both medical care process and outcome measures. Although it is appropriate to question the impact of certain common medical care practices on patient outcomes, there is a convincing body of medical literature and considerable medical consensus on the effectiveness of many medical care processes. Medical care process measures have certain advantages. They are frequently easier and less expensive to measure than are patient outcomes, and they are, for the most part, directly attributable to the actions of a health plan and its providers. On the other hand, outcome measures are probably more relevant to the concerns of consumers and purchasers; furthermore, they contribute to the knowledge base on medical care effectiveness. In summary, each of these approaches has merit, and there is the potential for synergy from the thoughtful combination of both types of measures.

Fourth, a performance measurement system should involve multiple data sources. The three most common sources of data are (1) administrative data sets, (2) medical records, and (3) enrollee surveys. Administrative data sets contain only limited information and are not uniformly available across the managed care industry, but when relevant information is available, the cost of retrieval is usually less than that of medical record retrieval, and the results apply to all enrollees rather than to a sample. Although medical records generally contain far more detailed information than do administrative data sets, the accuracy, completeness, and organization of medical records are variable. Enrollee surveys provide information frequently not contained elsewhere (e.g., information on enrollee expectations, satisfaction, and health outcomes). Surveys can be quite expensive, however, and there are often methodological issues of nonresponse and poor recall.

Fifth, there must be well defined mechanisms to ensure ongoing refinement and adaptation of a performance measurement system over time. Designating a specific set of performance measures to be used for reporting and monitoring purposes may cause health plans to focus much of their internal quality improvement efforts on the conditions selected. Furthermore, during the early years of such a system, the conditions chosen are likely to be those that are relatively easy to specify and measure. Refining the performance measurement system on an ongoing basis makes it possible to incorporate new conditions that offer the greatest potential for improved health status through improved health care delivery, and to remove existing conditions for which the costs of continued reporting are likely to outweigh the potential benefits.

TYPES OF PERFORMANCE MEASURES

In the development of performance measures, a good deal of work has been accomplished in the following areas: (1) condition-specific measures, including the development of a framework to guide the selection of condition-specific performance measures and the specification of valid and reliable measures; (2) measures of satisfaction; (3) general health status measures; and (4) system measures of access and use of service.[11]

Condition-Specific Measures

The RAND HMO Consortium, consisting of 12 health plans in collaboration with RAND, was formed to test the reliability and validity of quality-of-care measures that could be publicly released to compare health plans.[12] Based on estimates of the prevalence and health impact of various conditions, and the efficacy of available treatments, the Consortium identified 13 high-priority "target issues" for evaluating the quality of health care provided by various plans (Table 6–1). In identifying a set of conditions for health plan evaluation, the Consortium considered various practical issues, including (1) the availability and feasibility of collecting the necessary information on the quality of care, (2) the cost-effectiveness of the relevant improvement in care, and (3) the ability of the health plans being evaluated to influence the relevant improvements in quality of care.[13]

For most of the target issues listed in Table 6–1, there are numerous process and outcome measures that can be used to assess performance. Some

Table 6–1 High-Priority Topics for Quality Measurement

Target Issue	Proposed Measures
Prevention of low birth weight	Process measures to focus on the timeliness, frequency, and content of prenatal care
Childhood infectious disease	Immunization rates (by the age of 2) for measles, mumps, rubella, diphtheria, pertussis, tetanus, *Haemophilus influenzae b*
Treatment of otitis media	Process measures to focus on use of antibiotics for primary treatment and prophylaxis, and the use of ancillary diagnostic tests and surgery
Treatment of childhood asthma	Process measures to focus on timeliness and thoroughness of the treatment of asthma, with special attention to counseling and self-care behaviors; functional outcomes (e.g., lost school days) adjusted for disease duration, age, and race
Breast cancer early detection	Mammography rates among women over 50, stratified by age
Prevention of coronary artery disease	Rates of cholesterol screening by age and sex; adequacy of follow-up received for elevated cholesterol levels
Treatment of myocardial infarction	Process measures to focus on timeliness of treatment attention to recurrent angina, and appropriateness of medications
Treatment of diabetes mellitus	Process and intermediate outcome measures to focus on access, glucose monitoring, patient education, eye screening, and cardiovascular risk factor modification
Prevention of strokes	Process measures to focus on hypertension screening, appropriateness of treatment, and blood pressure monitoring; blood pressure results adjusted for age, sex, initial blood pressure, comorbidity and treatment duration
Treatment of hip fractures	Process measures to focus on the preoperative evaluation, prevention of complications, and postoperative care; outcomes adjusted for age, race, sex, fracture type, and comorbidity
Prevention of influenza	Immunization rates among those over 65, stratified by age and residence
Attention to the medical problems of the frail elderly	Process measures to focus on the appropriateness of medication regimens and attention to common geriatric problems (e.g., incontinence, confusion)
Overuse of surgical procedures and prevention of complications	Proportion of cardiac catheterizations, cholecystectomy, and hysterectomy, performed for indications rated appropriate; complication rates adjusted for age, sex, and comorbidity

Source: Adapted from Siu, A.L. et al., Choosing Quality of Care Measures Based on the Expected Impact of Improved Care on Health, *Health Service as Research*, Vol. 27, No. 5, pp. 629–631, with permission of the Hospital Research and Educational Trust, © 1992.

measures (e.g., use of medications in the treatment of myocardial infarction) require more complex assessment tools that rely on medical record audits or special data collection efforts, while others (e.g., mammography rates) may be estimated from administrative data sets available within many health plans.

For many of the target areas, performance measures already exist. For example, the RAND HMO Consortium has developed very comprehensive and sophisticated instruments for the assessment of prenatal care and the appropriateness of hysterectomies.[14-16] Certain managed care organizations have invested considerable resources in the specification of condition-specific performance measures, such as:

- newborn birth weight distribution
- premature delivery rate
- prenatal care visit rate
- chemical dependency recidivism rate
- appendectomy perforation rates
- pediatric asthma emergency department visits
- suicide rates
- influenza and pneumococcal immunization rates of the elderly
- admission rates for selected conditions (e.g., pediatric asthma, perforated or hemorrhaging peptic ulcer)
- breast cancer screening and follow-up
- cervical cancer screening and follow-up
- diabetic eye care[17-21]

InterStudy, the Health Outcomes Institute, and others have been pioneers in the development of the outcomes management system.[22] Data collection protocols (i.e., special questionnaires to be completed by patients or providers) either have already been developed or are currently being developed for a total of 28 conditions. Among these conditions are hypertension, hip fracture, angina, cataract, chronic obstructive pulmonary disease, hip replacement, low back pain, stroke, osteoarthritis of the knee, prostatism, and rheumatoid arthritis. Some measures have been subjected to feasibility testing and been found to be reasonably reliable and valid.[23]

Some of the projects funded by the Agency for Health Care Policy and Research (AHCPR) to translate practice guidelines into review criteria may yield useful information for the derivation of population-based performance measures. For example, the American Medical Review and Research Center is developing guideline-based review criteria for acute postoperative pain and urinary incontinence.[24]

In summary, the list of currently available performance measures specific to particular conditions is quite substantial, and there are some promising developmental projects under way.

Measures of Satisfaction

Enrollees are the ultimate consumers of health care services, and the long-term success of a health plan depends, in part, on its ability to meet the expectations of its enrollees. Enrollee surveys can provide information on overall satisfaction with the health plan, as well as satisfaction with various aspects of plan performance, such as (1) accessibility of the health plan, (2) satisfaction with the medical and nursing care provided, and (3) satisfaction with outcomes. Information obtained through enrollee surveys on certain aspects of performance may serve as early warning signs of plan difficulties (e.g., a measure of enrollee use of out-of-plan providers).

The majority of health plans survey their enrollees, and several well developed survey instruments are currently in use. In fact, standardization of enrollee surveys has been increasing in recent years. The Group Health Association of America (GHAA), in collaboration with Allyson Ross Davies and John Ware of the New England Medical Center, developed a survey instrument that has gained considerable acceptance within the managed care industry; some organizations use it in its original form while others use adapted versions.[25] More recently, the New England Medical Center, in conjunction with a consortium of employers, has developed the Employee Health Care Value Survey.[26] This survey not only incorporates many of the questions included in the GHAA survey, but also includes a health status assessment component and builds on a new collection of measures tested in the state of Iowa.[27] Other instruments have gained considerable acceptance in certain geographical areas, such as Minnesota,[28] the San Francisco Bay Area,[29] and Michigan.[30]

Unless health plans use a standardized instrument and survey process, the possibilities of comparisons across health plans are very limited. Differences in survey instruments (e.g., ordering of questions), survey questions (e.g., wording of questions, item response choices), approaches to conducting surveys (e.g., mail versus telephone) and approaches to sample selection will make comparisons difficult to interpret in a meaningful manner.

General Health Status Measures

The field of health status measurement has developed rapidly in recent years, and some measures may be useful for external reporting and assessment. In addition to the condition-specific measures discussed earlier, there is a 36-item short-form general health survey, known as the SF-36. Constructed to survey health status in the RAND Medical Outcomes Study, the instrument permits an assessment of eight health concepts:

1. limitations in physical activities because of health problems
2. limitations in social activities because of physical or emotional problems
3. limitations in usual role activities because of physical health problems
4. physical pain
5. general mental health
6. limitations in usual role activities because of emotional problems
7. vitality
8. general health perceptions[31]

The SF-36 and related instruments have been used extensively, and there is considerable evidence in support of their validity.[32,33] Extensive work is also underway to validate an abbreviated health status questionnaire, known as HSQ-12.[34]

General health status measures such as the SF-36 have been used to evaluate the effectiveness of alternative settings (e.g., the treatment of HMO patients through a special geriatric unit versus routine care)[35] and to assess the relative "health burden" of the enrolled populations of different health care plans.[36] These types of measures have generally not been used to monitor the quality of care provided by health plans,[37] however, and additional work is needed to establish the linkages between the overall health status changes in an enrolled population and a health plan's systems and processes for delivering care.

System Measures of Access and Use of Service

Various system measures may be useful in assessing certain aspects of a health plan's performance. Waiting times can provide a picture of health plan access, for example. A composite of health plan utilization measures, such as the following, is another mechanism for identifying possible barriers to access:

- High rates of emergency department use may indicate difficulties in obtaining access to primary care providers.

- Very low referral rates to specialists may indicate either excessive financial incentives to primary care providers to minimize referrals or perhaps a lack of qualified specialists within the network.
- Low use of mental health care providers and programs may indicate a failure to diagnose and treat mental health problems.
- A high proportion of enrollees having no contact with a health plan in a given year may indicate access problems.
- Disenrollment rates, especially voluntary disenrollment (i.e., not necessitated by factors such as change of employment or location), have long been used as a possible indicator of enrollee dissatisfaction with health plan service.

For the most part, these types of measures do not permit definitive conclusions about performance. Furthermore, because of differences in plan organization and benefit structure, patient mix, and local circumstances, caution is essential in making comparisons across health plans. In the same way that interrelated measures of financial performance (e.g., net worth, profit and loss statements, net income, cash flow) reveal an organization's financial status, a composite of utilization measures accompanied by benchmarking information can provide early warning signs of difficulties in health plans.

COMMON DATA SOURCES

At present, there are four major options for deriving information on health plan performance: (1) administrative data sets, (2) medical record audits, (3) enrollee surveys, and (4) special data collection efforts. There are some very promising projects under way to automate medical records and to build information infrastructures that link data from various transactions, such as ambulatory encounters, hospital episodes, medication orders, and laboratory tests.[38,39] Over the coming years, these efforts will have a significant impact on many application areas, including performance assessment. Only a small percentage of health plans are currently involved in these projects, however.

Administrative Data Sets

Most health care plans have some or all of the following types of administrative data sets:

- enrollment files that include patient identification, demographic information, and insurance data

- hospital episode data that include some or all of the standard billing data captured on common billing forms, such as the UB-92 used by the Health Care Financing Administration (HCFA) for Medicare billing
- encounter data that includes the type of visit, diagnosis, date of service, and other information captured on the common billing forms, such as the HCFA 1500 used by HCFA for Medicare billing
- ancillary services and pharmacy data

The majority of health plans currently have automated enrollment files and hospital episode data; fewer collect ambulatory encounter data in a systematic fashion. There is much variation in the collection of ancillary services and pharmacy data.

For the most part, administrative data sets have not been constructed for the purposes of quality assessment, and they are often limited in content, completeness, and accuracy. Health plans that compensate individual and institutional providers on a fee-for-service basis generally have more extensive administrative data sets, because the providers must complete and submit claims in order to obtain payment. Many group and staff models, and IPAs that compensate providers by means of a capitation payment (i.e., predetermined payment per enrollee), have only limited administrative data sets. Health plans that "bundle" payment for types of services (e.g., a single predetermined payment for all prenatal care during a pregnancy) are likely to have only limited information on the specific services provided (e.g., prenatal care visits). In many administrative data sets, the coding of clinical diagnoses and procedures is also problematic.

Administrative data sets are useful for (1) estimating certain performance measures and (2) drawing appropriate samples of cases for focused medical record audits. Many performance measures can be derived directly from administrative data sets commonly available in many health plans, such as mammography rates, childhood immunization rates, Papanicolaou smear rates, Caesarean section rates, and prenatal care visit rates.[40] In other instances, administrative data sets may be the starting point of a focused medical record audit, for example, to identify enrollees with diabetes in order to draw a sample of cases for detailed medical record review.

For health plans that do not collect ambulatory encounter data, automated data on diagnoses are generally available only for enrollees who have been hospitalized. This poses constraints on a plan's ability to monitor performance and underscores the need for the collection of a minimum, uniform encounter data set by all health plans.

Medical Record Audits

Most managed care organizations currently invest significant resources in the conduct of focused medical record audits. They would benefit greatly from improvements in the design of these focused audits (e.g., selection of appropriate sample sizes, use of appropriate statistical analysis techniques, standardized abstraction tools). Through the development of public domain study designs and instruments, it should be possible to achieve significant economies of scale, as well as to enhance the value of these audits.

Medical record audits are labor-intensive, and errors can easily occur during the abstraction process. In some instances, information may be missing from the medical record because of a failure to document or illegible entries. One study showed that the charts of only 50 percent of diabetic patients contained the results of their fundoscopic eye examinations.[41] Part of the reason for this situation may be that medical records are not always integrated across sites; all of the information required for a focused audit may not reside in a single medical record, and there may be no easy way to link one hospital's record with another or with a physician's office record.

Enrollee Surveys

As noted earlier, member satisfaction surveys are commonly used within the managed care industry, but to date the standardization of survey instruments and data collection methodologies has been limited. For the purposes of obtaining comparable data for public reporting and monitoring, there are two possible approaches: (1) encourage the common use of a comprehensive survey instrument and methodology, or (2) encourage the incorporation of a limited number of key survey questions into existing health plan surveys.

Health plans' use of a common survey instrument would have certain advantages. For example, detailed comparative information from patients regarding the medical care process and their satisfaction with many aspects of performance would be available. There would also be a potential to achieve certain economies of scale and to ensure uniformity in the conduct of the survey by arranging for a limited number of organizations with survey research expertise to carry out the survey (rather than having each health plan conduct its own).

An alternative approach is to identify a set of key questions that health plans may incorporate into various survey instruments. One advantage of this approach is that, although health plans may need more detailed information for internal quality improvement purposes, outside users interested in making comparisons across health plans may benefit from a more limited, but carefully selected, set of key performance measures. Standardization of a limited, core set of measures provides more flexibility for both innovation and the tailoring of surveys to meet local needs.

The standardization of only a few key questions poses certain methodological challenges. First, caution must be taken to ensure that the set of questions selected is not biased for or against particular types of health plans but rather provides a balanced picture of each type's strengths and weaknesses. Second, because the context of the questions can influence respondents' answers, it is important to consider how best to incorporate standard questions into commonly used surveys.

Special Data Collection Efforts

In certain instances, it may be desirable to rely on special data collection efforts. The use of special instruments that record data as care is being provided, in particular, has a great deal of potential both to enhance performance measurement and to assist physicians in medical decision making. If designed properly, these instruments afford a structured approach to provider decision-making processes and can give immediate feedback to providers on performance. Such concurrent data collection tools can sometimes be more cost-effective mechanisms for the collection of certain data than is retrospective medical record retrieval.

COMPARISONS OF HEALTH PLAN PERFORMANCE

Interest in establishing a "report card" on health plans' performance has surged in recent years, with the plans themselves, private purchasers, consumer groups, and, more recently, the public sector promoting the concept. The report card concept is the latest manifestation of what has been dubbed the "third revolution in health care," namely, the call for accountability.[42] An encouraging aspect of the efforts that have contributed significantly to shaping a framework for the development of a system of comparable performance measurement is their collaborative nature—most involve some combination of health plans, purchasers, labor, consumers, and the public sector.

InterStudy's Outcomes Management System

Ellwood and others associated with InterStudy have emphasized the need to measure "patient outcomes" and to establish a central mechanism for accumulating such comparative data.[43] At the core of InterStudy's outcomes management system is the SF-36,[44] the general health status measurement tool discussed earlier, which is used in conjunction with various condition-specific data collection protocols.[45]

There are currently two major collaborative efforts under way to collect comparative outcomes data from health care organizations by using outcomes management system measurement tools, known as TyPEs (i.e., Technology of Patient Experience).[46] The first, sponsored by the American Group Practice Association, involves single and multispecialty group practices in the collection of data on various conditions (e.g., angina, cataract development, diabetes) and procedures (e.g., hip replacement). The Managed Health Care Association, an association of major employers, is sponsoring the second effort. Working collaboratively with leading health care plans, the Association is collecting and pooling outcomes data on angina and asthma, and assessing the usefulness of such information for benefit plan management and quality improvement.

Building on the work of InterStudy, the Jackson Hole Group, a health care policy organization, has developed a model for a health outcomes accountability system in which a private sector board establishes standards for data collection and reporting by health plans.[47] Three data collection methods are proposed:

1. an annual sample survey of plan members to assess health behaviors, health status, and satisfaction with services
2. tabulations of selected provider activity records to document performance of desirable services, such as mammograms and immunizations
3. prospective collection of standardized clinical and outcomes data for selected conditions

The implementation schedule calls for gradually phasing in these reporting requirements.

Consortium Research on Indicators of System Performance (CRISP)

In 1992, the Center for Health System Studies, a research department within the Henry Ford Health System, began a study entitled Consortium Research on Indicators of System Performance (CRISP). Approximately 20 of the leading health care systems in the United States are actively involved in this initiative. They are testing and refining a set of performance indicators designed for a vertically integrated health care system responsible for providing health care to specific populations within a defined geographical area.[48]

The study is being conducted in two phases. The first phase focuses on the capabilities of current information systems, indicators presently in use, and the possibilities of producing additional indicators. The second phase involves developing, testing, and validating a common set of indicators for use as benchmarks. Measures currently being tested include hospital readmission rates and low-birth-weight incidence. Various condition-specific medical care process and outcome measures are in the early stages of development.[49]

The Massachusetts Purchaser/HMO Collaborative Effort

In 1990, a collaborative effort began between Digital Equipment Corporation, the Massachusetts Medicaid agency, and three area health plans (i.e., Fallon Community Health Plan, Harvard Community Health Plan, and Tufts Associated Health Plan) to develop and report on clinical indicators.[50] Six indicators were developed:

1. mental health inpatient readmissions and inpatient days per patient
2. Caesarean section rates
3. prenatal care in the first trimester
4. breast cancer screening rates
5. asthma inpatient admissions
6. high blood pressure screening

In 1992, each of the four participating health plans began to use the indicators. The comparative performance reports based on 1991 and 1992 data are among the earliest regionally based, purchaser–health plan efforts to assess health plan performance. This effort has subsequently been expanded to include many additional health plans and purchasers.

The Health Plan Employer Data and Information Set (HEDIS)

In late 1989, the HMO Group (an association of group and staff model HMOs) worked with four large employers (i.e., Bull HN Information Systems, Digital Equipment Corporation, GTE, Xerox) and Towers Perrin to establish a set of performance measures that would allow employers to assess the "value" of a health plan.[51] In September 1991, this effort culminated in the HMO Employer Data and Information Set (HEDIS 1.0), which included performance measures and reporting formats for quality, access and satisfaction, utilization, and financial data.

About this same time, another important effort was under way to identify utilization-based performance measures and to develop common reporting formats and specifications for use by health plans. The Minnesota health plans, in conjunction with the state government, established a Utilization Data Definitions Committee that, in 1992, released detailed specifications for extracting utilization data from common administrative data sets resident within most managed care organizations.[52] These detailed specifications are an essential step toward refining a methodology that will allow for true health plan comparisons.

Building on these two prior efforts, the National Committee for Quality Assurance, a national accreditation and quality oversight organization for managed care, established a Performance Assessment Committee (PAC) in October of 1992 to revise and refine HEDIS 1.0. The PAC membership included representatives of the original four employers as well as AEtna Health Plans, Harvard Community Health Plan, Kaiser Permanente, The Prudential Insurance Company, United Healthcare Corporation, and U.S. Healthcare. In addition, the PAC drew on the technical expertise of representatives from The HMO Group and the Minnesota Utilization Data Definitions Committee.

The effort has resulted in the release of The Health Plan Employer Data and Information Set, Version 2.0 (HEDIS 2.0) in November of 1993.[53] The HEDIS 2.0 core set of more than 60 measures covers five major areas of performance: (1) quality of care, (2) access and member satisfaction, (3) membership and utilization, (4) finance, and (5) descriptive information on health plan management and activities. Selection of measures was based on three criteria: relevance and value to the employer community, reasonable ability of health plans to develop and provide the requested data in the specified manner, and potential impact on improving the process of care delivery. Future iterations of HEDIS will contain refinements of current measures, the addition of new measures, and deletions of measures that do not provide useful data and information.

Medicare Performance Measurement Demonstration

In September 1993, the Health Care Financing Administration (HCFA) contracted with the Delmarva Foundation to identify an appropriate set of performance measures that peer review organizations can use to monitor the performance of HMOs with Medicare contracts.[54] This effort involved the review of more than 250 existing performance measures and resulted in the selection of three core measures (access to services, influenza immunization, and breast cancer screening) and two diagnostically-related measure sets (diabetes mellitus and ischemic heart disease). Currently, HCFA is pilot-testing these measures in several regions.

PUBLIC REPORTING OF PERFORMANCE DATA

The demands of purchasers and others for greater accountability and the subsequent collaborative efforts to develop common sets of performance measures have given impetus to the release of numerous "report cards" on health plan performance. In some instances, individual managed care organizations have issued report cards, while in other cases, report cards are part of a national or regional collaborative effort.

Early Efforts

Although the reporting of comparative performance information on hospitals is quite commonplace,[55] reporting such information on health plans is a relatively new phenomenon. The development of HEDIS serves as an ongoing catalyst for many health plan reporting initiatives.

The release of report cards by individual health plans has played a pivotal role in challenging others in the industry to release such information. In 1993, United HealthCare Corporation released a report card describing its health plans' performance, based on measures developed by the corporation's Center for Health Care Policy and Evaluation.[56] Shortly thereafter, both Kaiser Permanente Northern California and U.S. Healthcare released report cards based in part or entirely on HEDIS 2.0 specifications.[57-59] These groundbreaking efforts of individual health plans to measure and report their performance represent important steps toward making information on quality and access more readily available to guide decision making by consumers, purchasers, public sector representatives, and others. They are also tangible examples of the types of measures that may be included in a "report card" and the presentation of such information to consumers.

The report cards released by these leading health plans were only the beginning of a groundswell of public reporting efforts by the managed care industry. A very sizable proportion of managed care organizations responded to requests from employers for HEDIS 2.0 performance measurement data in 1994 and early 1995.

Several noteworthy community or state-based initiatives have also resulted in report cards. Building on the work of the Massachusetts purchaser/HMO group discussed earlier, the Massachusetts Healthcare Purchaser Group (a broad coalition of 23 purchasers) published comparative data for health plans in Massachusetts.[60] Working with a local business coalition in St. Louis, Missouri, HealthPages (a consumer publication) issued comparative data based on selected HEDIS 2.0 measures.[61]

In 1994, The National Committee for Quality Assurance also launched a national report card pilot project to test HEDIS 2.0 measures and to produce performance profiles for 21 health plans. This 1-year pilot project culminated with the release of a technical report in February 1995.[62] This project broke new ground with its emphasis on external auditing of health plan data. The 21 health plans participating in this project underwent an external audit to ascertain the strength of their internal data systems; and the accuracy and completeness of their source data sets and software programs used to generate HEDIS data.

Key Issues

The principle of public accountability through the reporting of performance data is well rooted in the managed care industry. There now appears to be cautious optimism that it is possible to compare health plans on the basis of cost and *quality*. At the same time, there are several concerns regarding the public reporting of performance data. These concerns relate to the limited set of performance measures currently available, the absence of adequate methods of adjusting for differences in the risk factors of health plans' enrolled populations, the status of health plans' internal data capabilities, the absence of external auditing functions, and limited knowledge of the information needs of consumers and other users of performance data.

Limited Set of Performance Measures

The types of performance measures currently being used in report cards do not represent a well balanced set. Most report cards place a great deal of emphasis on preventive services, and maternal and child care; there are fewer measures that address chronic conditions, mental health, and substance abuse. Specifying performance measures for the treatment of chronic conditions and mental health/substance abuse problems is particularly challenging because of the lack of clinical consensus regarding the appropriate care and the complexity of these conditions, the scarcity of practice guidelines, and the more extensive data requirements associated with performance measurement.

Performance measures currently available also tend to be medical care process indicators; only a limited number are outcome measures. Outcome measures tend to be more expensive to collect; many require special data collection instruments. Furthermore, it is difficult to attribute changes in outcomes to health plan interventions.

Risk Adjustment

One of the greatest concerns regarding the release of health care plan comparative data is the potential for unfair and unwarranted conclusions, owing to the lack of appropriate risk adjustment. For example, current report cards do not stratify health care plan results by socioeconomic status, even though certain conditions (e.g., asthma) are known to occur more frequently in lower socioeconomic groups, and those in lower socioeconomic groups are less likely to seek preventive services or early acute care interventions.[63,64] The data to adjust for many such risk factors are not readily available, the cost of collecting the information is significant, and there is a lack of consensus on how such information should be collected.

Health Plan's Data Capabilities

Thus far, the performance measures included in health plan report cards have been derived from the administrative data sets that many plans already have available. This is for two reasons. First, it is generally less expensive to produce performance measures from commonly available administrative data sets than to conduct medical record audits, to implement concurrent data collection mechanisms, or to carry out enrollee surveys. Second, administrative data sets allow a health plan to derive measures based on its entire enrolled population, rather than a sample, and to analyze results pertaining to various provider/patient groups (e.g., physician groups or individual hospitals). Analyses that can be "rolled down" to the individual provider level are more useful for quality improvement efforts.

Little information is currently available on the completeness and accuracy of information contained in health plans' administrative data sets. Unless supplemented with information from medical records, enrollee surveys, and other sources, however, administrative data sets are unlikely to provide a complete picture of the services/procedures received by a health plan's enrollees. A great deal of work is needed both to improve the completeness and accuracy of health plan administrative data, and to develop sound methodological approaches to the use of those data in combination with data obtained from other sources.

External Auditing

With the exception of the National Committee for Quality Assurance report card pilot project, the health plans' data presented in report cards released to date have not been subject to external verification by a third party. If consumers, purchasers, public representatives, and others are to rely on these data to select a health plan or to make purchasing decisions, it is imperative that an external auditing function be established. The external auditors should assess the integrity of the underlying data sets used by the health plans to generate performance measurement information and the adherence of the health plans to the specifications for the measures in the performance measurement set.

Information Needs of Consumers

The performance measures included in HEDIS, the Massachusetts Purchasers Coalition effort, and the Medicare demonstration project are ones that purchasers and/or health care professionals deem important to assess and feasible to implement at this time. The selection process has not been grounded in consumer research. Both research and evaluation are needed to ascertain how report cards may better serve the interests and needs of the lay public.

CONCLUSION

Comparative performance measurement and public reporting of health plan comparative data offer much promise. There is the potential to use the buying power and decisions of purchasers, labor, consumers, and public representatives to pressure the managed care industry to improve its performance continually. It is imperative, however, that adequate resources and attention be focused on addressing key concerns and developing the infrastructure necessary to produce reliable comparative data.

NOTES

1. Washington Business Group on Health, *Coalition Quality Initiatives: Twenty Case Studies* (Washington, D.C.: April 1992).
2. U.S. Congress, *Health Security Act*, Bill introduced before 103rd Congress, 1st session (Washington, D.C.: 1994).
3. See, for example, State of Maryland, Health General Article Section 19-501 et seq., 1993.
4. Group Health Association of America, *HMO Market Position Report* (Washington, D.C.: 1993).
5. Marion Merrell Dow, Inc., *Managed Care Digest: PPO Edition 1993* (Kansas City, Mo.: 1993).
6. J.E. Wennberg et al., The Need for Assessing the Outcome of Common Medical Practice, *Annual Review of Public Health* 1 (1980):277–295.
7. J.E. Wennberg and A. Gittelsohn, Health Care Delivery in Maine: Part I. Patterns of Use of Common Surgical Procedures, *Journal of the Maine Medical Association* 66, no. 5 (1975).
8. U.S. Public Health Service, *Healthy People 2000: National Health Promotion and Disease Prevention Objectives*, DHHS 91-05213 (Washington, D.C.: Government Printing Office, 1990).

9. Ibid., 521.
10. J.M. Corrigan and M.E. O'Kane, Assessing the Quality and Accessibility of Patient Care Provided by Health Plans (Paper prepared for the Physician Payment Review Commission, Washington, D.C., 1993).
11. Ibid.
12. A.L. Siu et al., Choosing Quality of Care Measures Based on the Expected Impact of Improved Care on Health, *Health Services Research* 27, no. 5 (1992):619–650.
13. A.L. Siu et al., A Fair Approach to Comparing Quality of Care, *Health Affairs* 10, no.1 (1991):62–75.
14. S.J. Bernstein et al., *A Literature Review of Indications for Non-Emergency, Non-Oncology Hysterectomy and Complications from This Procedure*, WD-5264-1-HF (Santa Monica, Calif.: RAND, October 1991).
15. P.J. Murata et al., *Prenatal Care: A Literature Review and Quality Assessment Criteria*, JR-05 (Santa Monica, Calif.: Rand HMO Quality of Care Consortium, 1992).
16. M.J. Sherwood et al., Medical Record Abstraction Form and Guidelines for Assessing the Quality of Prenatal Care, RAND Working Draft #6095-HF (Santa Monica, Calif.: RAND, June 1992).
17. L. Heinen et al., Quality Evaluation in a Managed Care System: Comparative Data To Assess Health Plan Performance, *Managed Care Quarterly* 1, no. 1 (1993):62–76.
18. L. Heinen and S. Leatherman, Quality Evaluation: A New State of the Art, *Group Practice Journal* (1992):38–43.
19. U.S. Quality Algorithms, *Performance Measures Manual* (Blue Bell, Pa.: 1993).
20. Kaiser Permanente Northern California Region, *1993 Quality Report Card* (Oakland, Calif.: 1993).
21. M.A. Bloomberg et al., Development of Clinical Indicators for Performance Measurement and Improvement: An HMO/Purchaser Collaborative Effort, *Joint Commission Journal on Quality Improvement* 30, no. 12 (1993):586–595.
22. InterStudy, Measurement and Management of Clinical Outcomes, *Quality Edge* 1, no. 2 (1992):1–71.
23. See, for example, D.M. Steinwachs et al., Feasibility of Outcomes Management System Implementation by Employers and Their Managed Care Organizations (Unpublished paper prepared for the Managed Health Care Association OMS Project Consortium, Johns Hopkins School of Hygiene and Public Health, Baltimore, November 13, 1992).
24. American Medical Review and Research Center, Develop, Apply, and Evaluate Review Criteria and Education Programs Based on Practice Guidelines, Phase I Report, Washington, D.C., April 1994.
25. A.R. Davies and J. Ware, *Consumer Satisfaction Survey and Users Manual*, Second Edition (Washington, D.C.: Group Health Association of America, May 1991).
26. H.M. Allen, The Employee Health Care Value Survey (Boston: The Health Institute, New England Medical Center Hospitals, 1993).
27. H.M. Allen, *Consumer Assessment of Health and Health Care: The Central Iowa Pilot Study* (Boston: New England Medical Center, 1993).
28. J. McGee, Minnesota Survey of Employees on Health Plans and Medical Care (Highland Park, Ill.: McGee and Associates, 1993).

29. Washington Business Group on Health, *Coalition Quality Initiatives*.
30. National Committee for Quality Assurance and J. McGee, Michigan Project Consumer Satisfaction Survey (Available from NCQA, Washington, D.C., or McGee and Associates, Highland Park, Ill., 1993).
31. J.E. Ware and C.D. Sherbourne, The MOS 36-Item Short-Form Health Survey (SF-36): 1. Conceptual Framework and Item Selection, *Medical Care* 30, no. 6 (1992):473–483.
32. A.L. Stewart and J.E. Ware, Jr., *Measuring Functioning and Well-Being: The Medical Outcomes Study Approach* (Durham, N.C.: Duke University Press, 1992).
33. C.A. McHorney et al., The Validity and Relative Precision of MOS Short- and Long-Form Health Status Scales and Dartmouth Co-op Charts: Results from the Medical Outcomes Study, *Medical Care* 30, no. 5 (1992):MS253–MS265.
34. Health Outcomes Institute, Update, *The Newsletter of the Health Outcomes Institute*, 2, no.1 (1995).
35. A.M. Epstein et al., Consultative Geriatric Assessment for Ambulatory Patients: A Randomized Trial in a Health Maintenance Organization, *Journal of the American Medical Association* 263 (1990):538.
36. Allen, *Consumer Assessment of Health and Health Care*.
37. S. Greenfield and E.C. Nelson, Recent Developments and Future Issues in the Use of Health Status Assessment Measures in Clinical Settings, *Medical Care* 30, no. 5 (1992): MS23–MS41.
38. The John A. Hartford Foundation, *Community Health Management Information System: General Overview* (New York: Benton International, 1992).
39. Workgroup for Electronic Data Interchange, *Report of the Workgroup for Electronic Data Interchange* (Available from Blue Cross and Blue Shield Association, Chicago, 1993).
40. S. Leatherman et al., Quality Screening and Management Using Claims Data in a Managed Care Setting, *QRB: Quality Review Bulletin* 17, no. 11 (1991):349–359.
41. W. Herman et al., An Approach to the Prevention of Blindness in Diabetics, *Diabetes Care* 6 (1983):608–613.
42. A. Relman, Assessment and Accountability: The Third Revolution in Health Care, *New England Journal of Medicine* 319 (1988):1221–1222.
43. P.M. Ellwood, Outcomes Management: A Technology of Patient Experience, *New England Journal of Medicine* 318 (1988):1549–1556.
44. Ware and Sherbourne, The MOS 36-Item Short-Form Health Survey.
45. InterStudy, Measurement and Management of Clinical Outcomes.
46. Ibid.
47. D. Lansky et al., Health Outcomes Accountability: Methods for Demonstrating and Improving Health Care Quality (Draft Discussion Paper, The Jackson Hole Group, Jackson Hole, Wyo.: March 16, 1993).
48. D.R. Nerenz and B.M. Zajac, Ray Woodham Visiting Fellowship Program Project Summary Report: Indicators of Performance for Vertically Integrated Health Systems (Chicago, Ill.: Hospital Research and Educational Trust, American Hospital Association, 1991).
49. D.R. Nerenz et al., Consortium Research on Indicators of System Performance (CRISP), *Joint Commission Journal on Quality Improvement* 30, no. 12 (1993):577–585.

50. Bloomberg et al., Development of Clinical Indicators.
51. The HMO Group, Kaiser Permanente, Towers Perrin, Bull HN Information Systems, Digital Equipment Corporation, GTE Corporation, Xerox Corporation, *HMO Employer Data and Information Set: Final Draft* (New Brunswick, N.J.: The HMO Group, 1991).
52. Utilization Data Definitions Committee, Minnesota Department of Human Services, Minnesota Department of Health, and Minnesota Health Plans, *Reporting Standards for Health Care Utilization Data* (Minneapolis, Minn.: February 1992).
53. National Committee for Quality Assurance, *Health Plan Employer Data and Information Set*, Version 2.0 (Washington, D.C.: November 1993).
54. Delmarva Foundation for Medical Care, *External Review Performance Measurement of Medicare HMOs/CMPs*, Draft Final Report, HCFA Contract 500-93-0021 (Easton, Md.: HCFA, June 1994).
55. U.S. Department of Health and Human Services, *Medicare Hospital Mortality Information* (Washington, D.C.: 1989).
56. Center for Health Care Policy and Evaluation, *A Report Card on Health Care: Performance Indicators for Evaluating a Health Care Delivery System* (Minnetonka, Minn.: United HealthCare Corporation, 1993).
57. G. Anders, Three HMOs Evaluate Themselves, *Wall Street Journal*, November 16, 1993, B-1.
58. Kaiser Permanente, *1993 Quality Report Card*.
59. U.S. Quality Algorithms, *U.S. Healthcare 1992 Quality Report Card* (Blue Bell, Pa.: 1993).
60. Massachusetts Healthcare Purchaser Group, *The Cost/Quality Challenge* (Boston: March 1994).
61. *HealthPages: A Consumer's Guide*, St. Louis Edition (New York: HealthPages, 1994).
62. National Committee for Quality Assurance, *Report Card Pilot Project Technical Report* (Washington, D.C.: February 1995).
63. L.I. Iezzoni, *Risk Adjustment for Measuring Health Care Outcomes* (Ann Arbor, Mich.: Health Administration Press, 1994).
64. G. Friday and P. Fireman, Morbidity and Mortality of Asthma, *Pediatric Clinics of North America* 35 (1988):1149–1162.

» Chapter 7 «

Closing the Loop: Are Clinical Pathways Our Most Comprehensive Approach To Measuring and Enhancing Performance?

Jonathan T. Lord

DESIGNING care properly is a key strategy for successfully meeting the challenges of the new health care environment. Succeeding in a world that is changing from a "supply side" system in which patients and the public had to accept what was offered in health care to a "demand-based" system that engages individuals as partners begins with a common sheet of paper. Visualized as a "stick figure" diagram related to the flow chart, the critical path delineates the sequential order, key factor analysis, and sites for completion of any process. A basic approach used in industry, particularly in engineering, critical paths are often complemented by other nonstatistical tools, such as cause-and-effect (Ishikawa) diagrams and systems diagrams. The fundamental difference between critical paths and systems diagrams is that critical paths are linear, simplistic depictions of process, whereas systems are circular in order to capture the influence of other processes/subprocesses on the process that is being described (Figure 7–1).

Several years ago, in conjunction with the application of continuous quality improvement theory to health care, a number of organizations began to apply the critical path approach to health care. These efforts started in hospitals and focused primarily on the services delivered to patients in a hospital setting. No one descriptive term has applied to this process, and names of similar activities have proliferated; critical paths or pathways, clinical pathways, care maps, clinical practice guidelines, and practice parameters are among the phrases used. In addition, the critical path process has often been implemented in the context of a "case management" program, a nursing management concept focused on delivering appropriate services to patients.

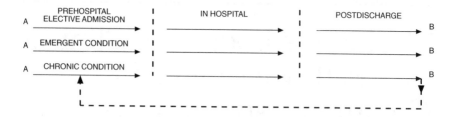

Figure 7-1 Linear and Circular Images of Care. *Source:* Reprinted from Lord, J., Practical Strategies for Implementing Continuous Quality Improvement, *Managed Care Quarterly*, Vol. 1, No. 2, p. 46, Aspen Publishers, Inc., © 1993.

Experience with the application of critical paths has been mixed. There is a clear relationship between (1) the integration of processes with people and (2) the ultimate success of the critical path. Factors for success include:

- linking the critical path to the organization's strategic plan
- integrating team members, including senior management, physicians, nursing staff, patients, and others outside the "organization" (e.g., payers, office managers, family members)
- designing care prospectively in all settings of care, including prehospital, in-hospital, and postdischarge
- engaging patients and their families, and preparing them for services, including wellness and physician strategies

Short routes to failure have been developing programs with a single focus (e.g., nursing), failing to integrate the critical path approach with existing quality programs, and viewing the creation of a pathway as an end point itself rather than as a program that progressively evolves.

STRATEGIC APPROACHES TO QUALITY

With the beginning of the era of quality improvement in health care, quality became a strategic function for organizations. An understanding of key products, key processes, and key quality characteristics, along with the renewed focus on customer-supplier relationships, has diverted attention away from a preoccupation with problems to a search for the best possible route and rationale for service. New market forces, such as the move to capitated payment schemes, have refocused efforts toward keeping people well rather than treating them only when they are ill.

At the core of the quality improvement theory is the notion that consistency in processes invariably increases the quality of products and services. In order to apply this theory, it is necessary to:

1. identify the core products of the organization (for health care providers, understand the needs of the communities served)
2. determine the key characteristics of the products or services that will meet or exceed the needs of the customers
3. design the clinical processes to deliver that service
4. redesign the business processes that support those clinical processes
5. apply the appropriate statistical process controls to measure and assess the performance of the process

The data set that is developed from these activities should tell a story about the value of service: clinical outcome, patient perception, and cost of service.

The focus of strategic efforts is the identification of the key processes that fulfill the needs of the community. Once those key clinical processes have been selected, the process of designing care can begin in earnest. The design effort has a number of elements that must be addressed:

- identifying the customer groups involved with the process
- defining the objectives and characteristics of the process
- establishing a framework for the formation of teams
- describing the process for design efforts

IMPORTANCE OF TEAM EFFORT

Identifying team members is a critical step in designing care. It begins with understanding stakeholders in the process—those individuals who actually touch patients, as well as the patients themselves. The "direct" stakeholder merits primary consideration for membership. Those who are outside the direct provision of services, but who have some vested interest in performance improvement should receive secondary consideration; involving individuals in those categories can create an "openness" in the system that will substantially contribute to process improvement.

Once constituted, the team should discuss ways in which the key process fits in the strategic direction of the institution, the goals of the team, and the mechanism for team activities. In addition, the team should identify required resources, such as educational research or benchmarking sources. Many teams establish a set of formal positions, including:

- leader
- facilitator
- recorder
- timekeeper

Some teams also ask for organizational "coaches" who have worked on continuous improvement or other team activities in the organization.

Other important preparatory functions include:

- compiling a historical data set
- gathering scientific literature pertinent to the process
- collecting benchmark data
- reviewing available care algorithms or pathways
- establishing clear-cut objectives

The team should then proceed with discussions about the key phases of care. Processes and subprocesses soon become obvious. Armed with this knowledge of existing processes, the team can begin to apply "what if" technology and explore the inverse of today's processes and alternates to existing processes. Through the exploration of new approaches, the team can begin to blend clinical process design with business process redesign. That activity, essentially re-engineering health care services, proves to be truly exciting for all team members.

The team effort is dependent on the supportive behavior of all team members. Engaging people in this effort requires an alignment of values, respect for the input of each team member, and quick and effective follow-up action to issues raised at team meetings. Fundamental to this effort is a sustained focus on patient care—not the process of designing care, continuous improvement, or team structure.

DESIGNING CARE*

The staff of the Anne Arundel Medical Center have been working with the development and implementation of clinical pathways since 1990. The pathway concept has been applied in a number of clinical areas as described in Figure 7–1. The efforts to build the clinical pathway for orthopedic patients typify the efforts of a pathway team.

Orthopedic surgery represents about 40 percent of the surgical caseload of the medical center, and the medical center is the third largest "pro-

---

Source: Reprinted from Lord, J., Practical Strategies for Implementing Continuous Quality Improvement, *Managed Care Quarterly*, Vol. 1, No. 2, pp. 45–50, Aspen Publishers, Inc., © 1993.

ducer" of joint replacements in the state of Maryland. It was for these reasons, as well as the size of the staff involved, that the medical center's first pathway team was formed.

The core members of the team (i.e., physicians, a nurse manager, the director of financial planning, physical therapy staff, and a representative of the management team) came together to review some basic performance information and to build consensus around the development of a clinical pathway. Data at the time showed that the average length of stay for patients was 12.2 days, and charges averaged $14,400 (not including surgeons' fees) for total hip replacements; patient satisfaction data were good, and the clinical staff were, by and large, satisfied with the services provided to patients. These data were compared to national and statewide normative data; performance was outside of the national norms at that time (average length of stay 10 days) but within statewide norms. Beginning with that meeting, and continuing with every subsequent meeting, the emphasis of the program was on building the best product for patients and the other customers involved. In fact, if there were any resource savings or elimination of "quality waste," the amount saved would be reinvested by the team in additional clinical services, equipment, or personnel needed for the provision of additional service to the community. This was incentive for the team to make resource decisions based on improvement in performance.

At that first meeting, the two groups of orthopedic surgeons were asked about differences in their practices (one group had an average length of stay of 10 days; the other group had a length of stay of 14 days). The answer was, "That's what we tell our patients prior to the time of surgery." That simple statement turned out to be one of the most important findings. Another function of the first meeting was to identify team members. A total of 28 key constituent groups were identified, ranging from the core group to support staff in social services, home health, dietary, and quality assessment, to patients and their families. The inclusion of patients and family members into the design team required the configuration of a focus group to help examine existing procedures as well as design efforts.

The team began meeting every other week; initial efforts were related to flowcharting existing ways of doing business, understanding the requirements of constituent groups, and learning about the treatment process itself. Early on, the visions of building the "best" service and of defining value of service (the triad of clinical outcome, patient and other customer perceptions, and cost of care) were placed on center stage as targets for the group. By the end of the third work month, the team had redesigned what began as an anecdotal 12.2-day process for the medical center into a

7-month process that began in the patient's home at least 30 days prior to the time of surgery and ended in the patient's home 6 months after surgery (Figure 7–2).

The innovations that resulted included the following:

- A common tool used to assess pain and mobility (a modified Harris Hip Scale) prior to and well after surgery in order to measure clinical outcomes.
- The patient beginning exercises prior to surgery.
- Physical therapy routines for the patient prior to surgery.
- Dietary counseling and meal selection prior to surgery.
- Discharge planning before the patient enters the hospital.
- Introduction to medical center staff (nurses, therapists, and support staff) prior to admission.
- Accomplishment of surgical case review prior to hospitalization.
- Development of "standard" orders requiring only one signature for the days of care.
- Home health follow-up for all patients and simplified admissions and discharge procedures.

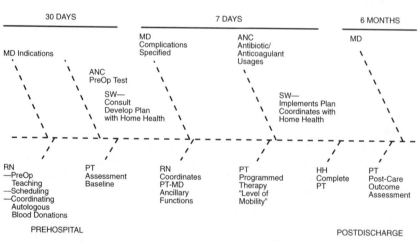

Figure 7–2 Clinical Pathway for Diagnosis-Related Group 209. *Source:* Reprinted with permission from *IQA-2: Continuous Performance Improvement Through Integrated Quality Assessments*, published by American Hospital Publishing, Inc., Copyright 1992.

During Month 4, the pathway was shared with members of the focus group. The focus group identified three areas for improvement:

1. Reduction of the number of times that the same historical information is elicited; one patient described being asked about an appendectomy 18 times.
2. The same-day admissions process.
3. Postsurgical exercise programs.

The team's solutions to these were as follows:

- Developing a simplified medical record that included an integrated patient database within the medical center's information system.
- Initiating a hotel stay program the night prior to admission.
- Beginning a postsurgical aquatic exercise program.

By Month 6, the first patient was registered to the pathway (Exhibit 7–1). In the intervening 6 months since the start of the team, that length of stay was reduced by 20 percent without making a single change in the system. As patients began using the pathway, a number of modifications were made. The first of these changes were made on an anecdotal basis; subsequent changes were made on the basis of data collected. Key treatment parameters that followed included clinical outcomes based on modified Harris Hip Scale measurements, perceptions of care based on feedback surveys that were part of get-well cards, and monitoring of the length of stay and charges for patients. Two years later, patient outcome scores show a 100 percent improvement on average, with the 80- to 85-year age group showing the most improvement (Figure 7–3). Patients were enthusiastic with the services provided (e.g., "My hospitalization was better than a cruise that I was just on!"). Practitioners were upbeat with an improved performance (e.g., length of stay, 6.1 days). There were also major surprises.

First, with the reduction in length of stay, the number of home health visits was expected to increase. In actuality, not only was length of stay decreased by one half, so was the number of postdischarge home health visits. The reason for this appeared to be the mindset and preparation of the patient prior to the time of surgery. Making the patient an "activist" in care was perhaps the single most important factor in changing resource consumption and influencing improved clinical outcomes.

Second, the charge to patients and payers was decreased by 30 percent (i.e., the actual charge took into account authorized rate increases by the state of Maryland), while the care and service provided to patients improved.

Exhibit 7–1 Standard Format Approach to Clinical Pathway for Diagnosis-Related Group 209

DRG 209: _____
EXP. LOS: _____
MD: _____
CASE MANAGER: _____
PATIENT SIGNATURE: _____
PROBLEM LIST: _____
_____
_____
_____

CLINICAL PATHWAY TOTAL HIP

		DATE:	Int	DATE:	Int
		Preop Evaluation		DAY 1	
Consults:		P.T. S.W. Dietary PRN		P.T. S.W.	
Tests:		Visit 1: EKG, CXR, Preop lab, Auto Blood Visit 2: Auto Blood, T&C		H&H (2hr Postop)	
PHYSICAL THERAPY TREATMENT:		See Home P.T. Evaluation Visit 2: P.T. visit —Evaluate Baseline Function —Instruct in Exercise, ambulation		Instruct: —breathing —quad sets —glut sets —ankle pumps —THR precautions As tolerates dangle, stand	
NURSING ACTIVITY: (Physical Immobility)				—Side to side —Reinforce exercise —Dangling as tolerated	
NURSING TREATMENT:				Post OP vs —> q4HR NV Cks q1HR × 8 —> q4HR Physical Assessment q8HR I&O/Pain Management IV IS q1HR Teds Off Q Shift Observe Drainage Reinforce Dressing PRN	

DATE:		DATE:		DATE:	
DAY 2	Int	DAY 3	Int	DAY 4	Int
Dietary PRN		O.T.		H.H./Home P.T.	
H&H, PT 7A		H&H, PT 7A		H&H, PT 7A	
Progress exercise: —assisted heel slide —assisted hip abd. Instruct: —bed mobility —dangle —stand —transfer training		Continue progress: —bed mobility —transfer —ambulation —exercise (add SAQs) To P.T. dept. if tolerates		To P.T. dept. —Exercise program —Ambulation program —Transfer training	
—OOB Hi chair × 3 —observe exercise × 3 —bed mobility		—OOB chair × 3 —observe exercise × 3 —bed mobility		—OOB chair × 3 —Ambulate × 2 —Observe exercise —Bed mobility	
VS q4HR NV Cks q8HR Physical Assess q8HR I&O D/C PCA —> Oral HL IS q2HR D/C Hemovac Assess Bowel Function		VS q4HR NV Cks q8HR Physical Assess q8HR D/C I&O/Oral pain med HL/IS q4HR Dressing change Assess Bowel Function		VS q8HR NV Cks q8HR Physical Assess q8HR D/C HL IS q4HR Fleet's Enema PRN Oral pain med Assess Bowel Function	

continues

Exhibit 7–1 continued

	DATE:		DATE:	
	Preop Evaluation	Int	DAY 1	Int
TEACHING: (Knowledge Deficit)	Visit I: Preop teaching —Orient to unit —View THR video —Clinical Pathway —IS/C&DB —Pain Management —Equipment/ Treatments		Review IS —Positioning Do's & Don'ts —Explain VAS —Explain reason for blood transfusion	
MEDICATIONS: (Pain)	Consult with Anesthesia for Medication Taken Morning of Surgery		Preop antibiotics Iron TID Coumadin HS MOM 30cc PRN HS po Colace 100mg bid Epidural/PCA	
NUTRITION:	Visit 1—Complete preadmission nutrition questionnaire. Nutrition Evaluation		NPO-Preop DAT ___	
DISCHARGE PLANNING: (Impaired Home Health Management)	Visit 1—SW evaluation —Living Situation —Relatives —Source of Income —Ability to Function —Pt. Expectations Home Health Evaluation —Safety —Equipment		Identify NH placement/Inhome caregiver PRN LTMA Application PRN	
KEY PATIENT OUTCOMES:	Key Patient Outcomes Pt. will: 1. State 　—unit/SDA routine 　—use of VAS 0–10 scale 2. Demonstrate 　—C&DB/IS		*Demonstrate use of:* —IS/C&DB q1hr —Abduction pillow and OFT q1hr *State:* —VAS 0–10 less than 5 —Activity restrictions —Rationale for transfusion	
ADDITIONAL PATIENT OUTCOMES:				

Source: Reprinted from Lord, J., Practical Strategies for Implementing Continuous Quality Improvement, *Managed Care Quarterly*, Vol. 1, No. 2, pp. 48–49, Aspen Publishers, Inc., © 1993.

DATE:		DATE:		DATE:	
DAY 2	Int	DAY 3	Int	DAY 4	Int
Review: TEDS Application/ rationale Medication Pathway Progress		Review: Pathway Progress		Review equipment available for home use	
Antibiotics Iron TID Coumadin HS MOM 30cc HS PRN Colace 100mg bid Mylanta 30cc q4HR PRN Oral pain meds		D/C Antibiotics Iron TID Coumadin HS MOM 30cc HS PRN Colace 100mg bid Dalmane 30mg HS PRN Mylanta 30cc q4HR PRN Oral Pain Med		Iron TID Coumadin HS MOM 30cc HS PRN Colace 100mg bid Mylanta 30cc q4HR PRN Dalmane 30mg HS PRN Oral Pain Med Fleets Enema PRN	
50% Meal consumed		75% Meal consumed			
Level of Care if NH needed ID Screen LTMA application completed PRN Inhome caregiver PRN		NH Inquires LOG Progress Inhome caregiver PRN		Identify equipment needs HH PT Ordered Coordinator meets with family/patient, RN, HHA assigned Obtain patient services	
Demonstrate use of: —TEDS *State:* —Medication management —Review clinical pathway				*State:* —Discharge —Equipment needs	

118   THE EPIDEMIOLOGY OF QUALITY

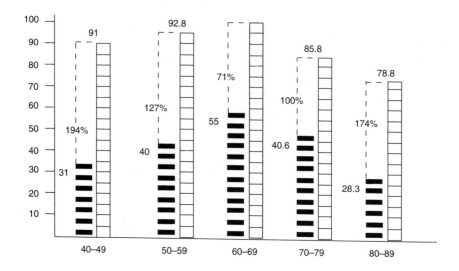

Figure 7–3 Harris Hip Preoperative and Postoperative Scores (July 1–December 31, 1991). *Source:* Reprinted from Lord, J., Practical Strategies for Implementing Continuous Quality Improvement, *Managed Care Quarterly*, Vol. 1, No. 2, p. 50, Aspen Publishers, Inc., © 1993.

This "story" was repeated in virtually every service line using this approach. Commonalities of the process include:

- *Educating, orienting, and "activating" the patient to become a part of the process.* To this end, mental health professionals have become an important part of every team.
- *Allowing the teams the freedom to explore all treatment activities through a focused effort to "reverse assumptions" and benchmark against "best" practices.*
- *Emphasizing patient care, not the process of continuous improvement, as the reason for the team's existence.* This philosophy was buttressed with "just-in-time" applications of the scientific tools of continuous improvement.
- *Concentrating on the systems approach as opposed to an episodic approach to care.*
- *Relying on information systems to reduce redundancy, eliminate rework, and aggregate data about performance.*

MEASURING AND ASSESSING PERFORMANCE

The development of performance indicators is integral to the design of the pathway of care. The indicators should originate in the objectives that were justified by key customer groups and should reflect the value of service. Thus, the data set should depict:

- outcome
- perception of service
- cost
- key process variables

Other chapters in this book describe the development and application of individual-based monitoring systems.

CONCLUSION

Team design of care and the application of clinical pathways (or algorithms for treatment) are exciting and fundamental strategies in contemporary quality programs. They are based on a blend of transitional clinical approaches to the provision of care, team concepts, and statistical process derived from continuous improvement theory and process management principles adapted from systems engineering. Taken together, this synergistic approach will lead to innovative improvement in the quality of care.

» Part II «

Reports from the Field

» Chapter 8 «

Hospital Use of Clinical and Organizational Performance Indicators

Anne M. Warwick, Anita M. Langford, and Judy A. Reitz

CONSUMERS want to know whether they are getting the most value possible for their health care dollar. The notion of value, or the worth of a health service compared to its price, underlies the new attention that consumers and payers are giving to the dimensions of cost and quality; they expect a certain level of quality for the price paid. Thus, consumers are holding health care providers accountable for performance. Outcome-based provider performance data are gaining prominence in the decision-making process for all purchasers of health care. In addition, performance data are likely to play a major role in health care reform.

Those in the health care professions have long been concerned about quality. Doing the "right thing" is the very nature of medicine. Medical knowledge has increased because physicians, nurses, and other providers observed patient outcomes, saw problems, and sought solutions. Indeed, such discoveries as antibiotics have often resulted from the quest to improve health and increase longevity. The public health gains in the early part of this century occurred largely because individuals asked the right questions and methodologically pursued answers to those questions.

Part of the quest toward quality resulted in the creation of the Hospital Standardization Program in the early 1900s. The forerunner of the Joint Commission on Accreditation of Healthcare Organizations, this program focused on measuring outcomes of hospital surgical patients. Unfortunately, when it was folded into the Joint Commission in 1951, compliance with standards became the measure for quality. It was assumed that if a hospital complied with standards, outcomes would automatically be good. In the first decades of the Joint Commission, standards mandated that hospitals evaluate "optimal" care through quality assurance activities such as audits and medical record reviews. These early studies—morbidity and mortality reviews, for example—were retrospective, time-limited, and conducted primarily to identify problems. More recently, systematic monitoring and evaluation of important aspects of patient care have replaced the problem-focused approach. By continuously collecting data and taking actions to improve care, providers expect to ensure quality in a more comprehensive way.

The early efforts to quantify quality focused predominantly on clinical outcome measures on the assumption that quality is a fixed target. An outcome refers to the patient's experience as a result of an activity being performed or, conversely, as a result of an activity not being performed. It is the end product. Death, postoperative wound infection, and low birth weight are common clinical outcomes. The cause-and-effect nature of rates was rarely examined comprehensively in initial quality assurance efforts. A less than optimal rate was usually deemed a problem with individual caregiver performance or skill.

The continuous quality improvement approach builds on this quality assurance model and broadens its scope. The Japanese used the methods of continuous quality improvement in the automobile and audiovisual equipment industries after World War II. Deming and Juran, who described these methods, were ultimately responsible for the well-known Japanese success story that ensued. Continuous quality improvement is based on the concept that it is possible to improve quality by continually refining processes rather than by simply removing problem workers. Its techniques focus on an understanding of complex processes, goal-directed sequences of related activities or steps that result in an end product or outcome. The use of statistical methods and removal of barriers to effective communication are other hallmarks of continuous quality improvement.

As U.S. industry began to understand and use continuous quality improvement methods, those concerned with quality in health care began to assess the usefulness of these methods in health care. The National Demonstration Project on Quality Improvement in Health Care, which began in 1987 and paired quality experts from industry with health care groups, demonstrated a high rate of success in improved care, better services, and cost savings.

The movement from quality assurance to continuous quality improvement is an inevitable step in the evolution of health care quality management. Not only is giving the best care possible the moral position that hospitals should take, but also gaining a reputation for quality makes good business sense in the present highly competitive health care environment. A hospital's very survival depends on its ability to demonstrate optimal patient outcomes at a fair and reasonable cost.

The relationship between performance of process and results is well established in health care, as it is in other industries. Performance also affects cost. Poor performance can result in expensive errors, duplication, and costly rework. Thus, inherent in the ability to improve continuously is the need to measure and evaluate performance. Continuous data collection in key areas and analysis of performance indicators help hospitals identify opportunities for improvement. Performance indicators are quantitative measures of clinical, support, or financial functions that, among other factors, affect patient outcomes. Although not direct measures of quality, performance indicators can reveal a great deal about quality.

VIEWS OF QUALITY

The fact that judgments about quality are frequently subjective and based on the parochial needs and expectations of the hospital's customers—both internal and external—complicates a hospital's quest to achieve optimal quality. Whether or not an outcome is considered optimal depends on who is making that determination. An accident victim who experiences a residual functional deficit while expecting a full recovery is unlikely to perceive this outcome as optimal. Even if the functional level achieved is far greater than the functional level reasonably predicted, the patient's perception of the quality of care is unlikely to change. A health care practitioner's viewpoint, on the other hand, may be just the opposite. Alternatively, a third-party payer may weigh the benefits of the patient's higher than predicted functional level against the cost of an extended length of stay or greater than average ancillary services charges.

Different groups of customers base their judgments about quality on different elements:

- Patients
  1. level of communication, concern, and courtesy of the providers
  2. degree of symptom relief (e.g., pain relief)
  3. level of functional improvement/maintenance (e.g., ability to use limb after a stroke)
  4. level of competence demonstrated by practitioner/employee knowledge or skill
- Practitioners
  1. state-of-the-art technology
  2. freedom to act in full interest of the patient

- Purchasers
  1. efficient use of resources
  2. appropriate use of resources
  3. optimal patient outcomes

Performance data that are accurate and relevant can assist a hospital in making the transition from subjective judgments about quality based on customer perceptions to objective measures of specific processes related to customers' needs.

PROCESS VS. OUTCOME INDICATORS

Performance indicators are useful in assessing either a process or an outcome. Despite the reliance of continuous quality improvement programs on processes and a systems approach to identify improvement opportunities, outcome indicators remain a valuable source of information. They offer immediate feedback to purchasers, as well as to providers of health care services. Patients who are seeking services are demanding information about a facility's or practitioner's results. Likewise, third-party payers and employers are making business decisions based on outcome measures.

Outcome indicators are particularly useful for improving care when they direct caregivers to the processes that contribute to the outcomes. They also help establish priorities by highlighting processes that are particularly troublesome and contribute to poor outcomes. In addition, this type of indicator is useful for determining the effects of changes made in processes. Outcome measures can be misleading, however. There may be variables beyond the control of the facility that influence outcomes. For example, the patient's age, severity of illness, socioeconomic situation, and co-morbid conditions all potentially influence clinical outcome.

A process indicator focuses on the activity of giving patients care, either directly or indirectly. Patient care functions, such as clinical assessments and specific interventions, can be process indicators. For example, the length of time that patients must wait in the emergency department for triage is a process indicator. When there is evidence that the process of care is linked with the outcome, process indicators are appropriate. In this example, there is good reason to believe that a quick triage assessment will improve patient outcomes by ensuring immediate care to those who need it. Process indicators are also useful when outcomes are difficult to evaluate, when they occur in the future, or when the patient has moved to another site of care. The best clinical process indicators are those for which a scientific basis in the literature exists to link them to clinical outcomes. Without that, the consensus of experts is essential.

Process and outcome indicators can measure both desirable and undesirable occurrences. They can also be designed to examine particular characteristics of quality (e.g., effectiveness, efficacy, appropriateness, continuity, safety, efficiency, availability, timeliness, patient perception of care). Furthermore, they can be used to assess nonclinical activities. In fact, to get a full picture of a health care organization's performance and to ensure continuous improvement, both process and outcome indicators of key clinical and nonclinical organizational activities are necessary (Exhibit 8-1).

PERFORMANCE INDICATOR DEVELOPMENT

An interdepartmental or cross-disciplinary approach to the development of indicators is most effective in helping the organization evaluate all relevant aspects of the delivery of patient care. Ideally, a group of "experts" (i.e., individuals familiar with the processes that occur at the grassroots level) should develop the indicators by using a consensus decision-making approach and quality management techniques. This method has the advantage of encouraging acceptance of the indicators developed, as well as the quality improvement process itself. The members of this group should possess a basic understanding of the organization and its day-to-day work, but they should also fully appreciate its inner character, philosophy, strengths and weaknesses, and its vision for the future. The group's level of expertise in particular areas, knowledge of the hospital functions, and optimal clinical outcomes, enhance the face validity and relevance of the indicators selected. Knowledge of statistical methods of measurement is also important to the successful development of measurable indicators. See Chapters 2 and 3 for information on the reliability and validity of indicators.

The process of identifying meaningful performance indicators must begin with a delineation of the care and services provided by the organization as a whole and by individual departments, as well as with an assessment of customer needs and expectations in relation to provider services. Not only the key procedures and treatments, but also all the other activities that make up the entire range of functions within the organization should be listed. In addition, listing the roles and responsibilities of personnel, characteristics of patients served, typical conditions and diagnostic categories treated, and specialized services performed can result in a comprehensive description of the facility's scope of care. Quality manage-

Exhibit 8–1 Indicators

	Outcomes	Processes
FINANCIAL	Collection rate Days of revenue in accounts receivable Budget variance	Insurance verification Preauthorization of admissions Information system data entry Supplies/equipment selection and usage Workers' compensation claims follow-up
CLINICAL	Adverse drug reactions Patient satisfaction Readmission within 72 hours of discharge for same problem Caesarean section rate	Patient examination Medication administration Meal tray preparation and delivery Patient/family education
ORGANIZATIONAL	Regulatory body citation Delayed discharge Procedure delay Staff turnover rate/vacancies	Product selection Preventive maintenance Results reporting Patient registration Sequence of test scheduling Recruitment procedures

ment techniques useful for this exercise include brainstorming, cause-and-effect diagrams (fishbone), task listing, flowcharting, and multivoting.

The next step in developing meaningful indicators is to determine the important aspects of care from the key functions listed in the first exercise. A function is a series of interrelated processes that are targeted to a specific goal. Although usually delineated by departments or units, functions often overlap in spite of the boundaries imposed by the organizational structure. For example, physicians, nurses, pharmacists, technicians, and clerks may all be involved in ordering medication.

Functions may be clinical, organizational, or financial in nature. Almost all are interrelated, and most contribute to patient outcomes directly or indirectly. Clearly, inaccurate testing of blood samples may result in a missed diagnosis or inappropriate treatment, but not all connections to patient outcomes are so obvious. A hospital's poor accounting practices may result in a higher than expected postoperative wound infection rate, for example, because the hospital cannot afford the staffing level needed to care adequately for postsurgical patients.

Selection of Most Useful Indicators

The number of processes performed in a hospital is vast. It would be impossible and unnecessary to measure them all. Therefore, hospitals should monitor those indicators that will provide the most information and the best opportunity to improve performance. One way to maximize the time and resources spent on quality improvement activities is to focus on high-volume, high-risk, problem-prone, and high-cost issues.

High-volume functions are direct patient care activities that are done very frequently and affect a large proportion of patients. Procedures and/or particular diagnostic categories may also fall into the high-volume category. A hospital may choose to examine many aspects of a surgical procedure that is its specialty. Clinical outcome of the procedure is one obvious aspect of interest, but examining the interdisciplinary functions that are related to this procedure is important as well. Perioperative care, staff skill level, accuracy, and timeliness of medication administration are related processes that may affect the surgical patients' outcomes, for example.

High-risk functions are those that, if performed incorrectly, have the greatest chance of detrimental effects on the patient or the organization. Functions may also be high-risk by their very nature. For instance, routine surgery performed perfectly on a patient with several complicating conditions becomes high-risk surgery. In addition, complex, untested procedures can pose high risks, even if done flawlessly. A procedure or service that is newly offered at the facility should be included in this category.

Problem-prone activities are those that have historically caused problems for the facility or for patients. Generally, these are activities that require a significant degree of interdepartmental coordination. Complaints and areas of dissatisfaction most likely to alienate existing or new customers should be monitored and evaluated as problem-prone activities.

Functions that reflect high cost, as well as those directly related to the organizational strategic objectives, should be included in the monitoring and evaluation activities.

Following are possible sources of data to help isolate high-volume, high-risk, problem-prone, or high-cost activities:

- High volume
  1. operating room utilization
  2. case mix analysis
  3. outpatient clinic visits/volume
  4. ancillary usage by service reports
- High risk/problem prone
  1. risk management reports
  2. patient complaints
  3. claims reports
  4. customer satisfaction survey
- High cost
  1. collection rate
  2. competitive position
  3. outliers
  4. denied day reports
  5. payer mix
  6. length of stay analysis
  7. bad debt ratio
  8. budget variance reports
  9. cost per discharge

Weaknesses in an organization's information system may become glaringly obvious during the selection of useful indicators and may, in fact, point to a significant opportunity for improving the quality of available data.

Reliability and Validity

Performance indicators must be reliable and valid in order to be useful as a measure of quality. Reliability refers to an indicator's stability, consistency, and accuracy in the identification of occurrences. Testing for reliability focuses on the accuracy and completeness of indicator occurrence identification to minimize the possibility of false-positives, which tend to increase indicator rates inappropriately, or false-negatives, which fail to identify true occurrences, thereby making the rate appear artificially low. Either situation adversely affects the organizational response to the indicator rate.

Validity refers to the extent to which an indicator actually measures what it purports to measure. The two most common types of validity are content and face validity. Both types have no empirical basis, but rely on the judgment of individuals who are experts. In order for indicators to be valid and relevant, there must be an opportunity to improve care. An indicator that identifies occurrences related primarily to the severity of illness would lack validity, because there is no opportunity to influence outcomes by improving care. An indicator that lacks reliability will also lack validity; the reverse is not necessarily true, however.

Sentinel vs. Rate-Based Indicators

There must be an effort to resist the inclination to focus on sentinel event indicators. A sentinel event is an unexpected, usually avoidable event that must be investigated because the consequence is so grave to the patient or the organization. Events that are often considered sentinel event indicators include:

- maternal death
- adverse drug reactions
- perioperative death
- blood transfusion reactions
- neonatal death

Because sentinel events are typically rare and isolated instances, their usefulness in a systematic assessment of adequacy of care is limited. A process analysis based only on sentinel events yields insufficient opportunities for improvement. Nonetheless, sentinel events can be an important source of information. For example, the case analysis of a perioperative death may reveal that preoperative electrocardiograms are not always placed in the patient's medical record prior to surgery because written reports are delivered only on certain days of the week. At all other times, staff transcribe a telephone report into the record. The potential sources of error in this process are numerous, but may never become obvious until an unfortunate event occurs.

Unlike sentinel event indicators, which describe single occurrences, rate-based indicators aggregate data of many events over a specific time frame. Rate-based indicators express the ratio or proportion of occurrence of the event being measured in relation to the overall population at risk for the occurrence. To determine the rate, the number of occurrences (the numerator) is divided by the number of total patients at risk for the occurrence (the denominator). There is a built-in assumption that, even when care is optimal, the event will occur at a certain level.

Indicator Statements

Indicators are often expressed as a single phrase, such as postoperative wound infections or inpatient mortality. In order to be most useful, however, an indicator must be clearly, precisely, and concisely defined. Fundamental information about an indicator is necessary to enhance its usefulness. For example, a written indicator statement should describe the indicator's intended focus. Terms used in the indicator statement must be fully explained so that everyone involved with data collection or analysis understands them in exactly the same way. The indicator should be identified as a process or outcome, rate-based or sentinel event type. The rationale for use and supporting evidence should be stated, and the population expected to be affected should be described. Data sources and the logistics of gathering data elements must also be explained. Finally, possible reasons for variation in the data should be explored. Exhibit 8–2 is an example of an indicator worksheet for a common rate-based outcome indicator: Postoperative wound infection.

INDICATOR DATA EVALUATION

Use of Data Thresholds

Indicator threshold rates are critical in the evaluation of indicator data, as they trigger the organizational response. When the indicator reaches or exceeds the threshold rate, further examination of the process assessed by the indicator must take place.

The threshold for a sentinel event indicator is always set at zero, because every such incident should be investigated. For other types of indicators, threshold rates may be determined in various ways. The consensus opinion of the content experts charged with selecting and defining the organization's indicators of key aspects of care may determine the threshold rate; a literature review and external sources considered authorities should support this opinion. In the example of the postoperative wound infection (see Exhibit 8–2), rates are to be compared to the operation-spe-

Exhibit 8–2 Indicator Worksheet

Postoperative Wound Infection

A. *Definition of Terms*
   1. Postoperative wound infection: See Centers for Disease Control and Prevention (CDC) definition.
   2. Class of operation: range from Class I–IV: See CDC definition.
   3. American Society of Anesthesiologists (ASA) rating: This is a rating of the physical status of the patient going into surgery ranging from 1 (healthy patient with no significant physical problems) to 5 (patient with severe underlying disease, in poor physical condition).
   4. The duration of operation as compared to a procedure-specific standard: The CDC has compiled a list of operations and calculated the range of times for each operation. They have determined the 75th percentile for each operation. For example, 75% of herniorrhaphies take 2 hours or less. Any herniorrhaphy taking over 2 hours exceeds the 75th percentile.

B. *Type of Indicator*
   1. Process _____   1. Sentinel event _____
   2. Outcome _____   2. Rate-based _____

C. *Rationale for Selection*
   1. Postoperative wound infection represents a patient outcome that may suggest the need for further review. A national study done by the CDC (Scenic Study) revealed that calculation of accurately compiled and analyzed surgeon-specific wound infection rates was associated with a 20%–38% decline in infection rate. Subsequently, in 1987, the Surgical Infection Society adopted a standard on wound surveillance for infection, stating the surgeon-specific wound infection by risk class should be determined and reported in all hospitals conducting 2,000 or more operations/year.

   Wound infection rates stratified by risk factors are appropriate for comparison with national rates and among surgeons.

   2. Selected References:
   Nosocomial Infections in Surgical Patients in the United States, January 1986–June 1992.
   Nosocomial Infection Rates for Interhospital Comparison: Limitations and Possible Solutions. CDC.
   3. Compiling surgeon-specific infection rates does not imply that the surgeon alone is responsible for wound infections. It is recognized that other members of the surgical team, the environment, supplies and materials, as well as procedures, can influence infection. However, an increase in adjusted wound infection rate relative to the national rate (or relative to an individual surgeon's previous rate) indicates that a problem may be present that warrants examination of the entire surgical process. It is critical to enlist the active participation of the surgeon, as well as others, in this examination.

D. *Description of Indicator Population*
   1. Subcategories: Patients having Class I or II operations will be stratified into three risk groups:
      0 risk factors: low risk

continues

Exhibit 8–2 continued

>   1 risk factor: medium risk
>   2 risk factors: high risk
>   2. Data format
>   a. *Numerator:* wounds that developed postoperative wound infection
>   b. *denominator:* number of patients in risk strata
> E. *Data Elements*   *Data Sources*
>
> | Data Elements | Data Sources |
> |---|---|
> | 1. Surgical operation performed | Operating room (OR) computer report |
> | 2. Class of operation | OR computer report |
> | 3. ASA rating | OR nurses backup |
> | 4. Length of time of operation | Time computed from OR computer report |
> | 5. Table of percentiles of operation length to compare length of each operation | OR computer report, CDC tables |
> | 6. Wound infections | Questionnaires sent to individual surgeons, returned to infection control |
> | | Review of daily contact list for re-admits, or visits to emergency department for wound infection |
> | 7. Validation of surgical wound infections reported by surgeon or detected on daily contact | Patient chart, discussion with surgeon, microbial data |
>
> F. *Underlying Factors*
>   1. Patient-based factors
>   a. Severity of illness: very important. ASA rating shown to be a good indicator of patient's physical status relating to risk of infection.
>   b. Co-morbid conditions: very important. ASA rating incorporates co-morbid conditions, making it a good overall indicator of patient condition.
>   c. Other patient factors: not very important.
>   d. Risk adjustment factors: see above.
>   2. Non–patient-based factors
>   a. Length of operation: very important.
> G. *Existing Databases*
>   The CDC publishes operation-specific rates to which adjusted indicator-specific comparisons can be made. Accurate comparisons will depend on adequate sample sizes. Confidence limits will be calculated for each rate as a gauge of the statistical power of the analysis.

cific rates published by the Centers for Disease Control. Furthermore, thresholds differ according to severity of illness and other risk factors.

Figure 8–1 demonstrates how exceeding the nationally acceptable rate for Foley catheter–associated urinary tract infection motivated Hospital XYZ to examine its process for Foley catheter insertion and maintenance. Two activities in the process were identified as opportunities for improve-

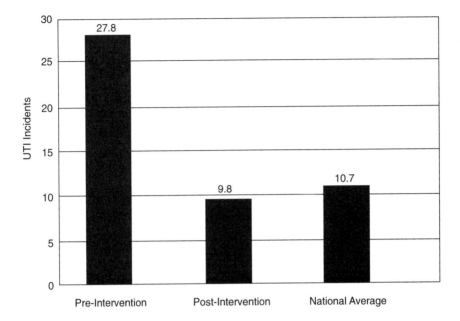

Figure 8–1 Hospital XYZ Foley Catheter–Associated Urinary Tract Infections (UTIs)

ment: daily catheter care and emptying of the receptacle bag. After changes were made in the process, the infection rate dropped below the national rate.

A threshold may also be based on the organization's own internal objectives. Certain rates may be targeted because of an expected program expansion or because of a desire to exceed a generally acceptable rate, for example.

Statistically derived numerical levels, preferably obtained from a large database are sometimes the source of thresholds. Because an individual hospital is unlikely to have enough homogeneous groups of patients or the resources required to establish a meaningful database of its own, hospitals may pool their indicator data. Pooling permits the progressive accumulation of large volumes of data and calculation of the statistical average and a range around the average.

The statistical average or mean of this large data set can be used to determine the baseline or usual rate for the occurrence of the event of interest. A more valuable statistic—because it describes the range above and below the average—is the standard deviation. A threshold rate is typically established at one or two standard deviations from the mean. Comparison with data from external aggregate databases is particular useful for rate-based clinical indicators such as:

- inpatient mortality
- Caesarean section
- vaginal birth after Caesarean section (VBAC)
- postoperative wound infections
- hospital-acquired infections
- readmissions to acute hospital within 72 hours

Comparative database reports should be systematically examined in relation to the hospital's expected performance (threshold) and in relation to rates reported by other participants in the database. Questions such as the following should be considered:

- How do the hospital's rates compare to the mean?
- How do they compare to the internally established threshold?
- How many standard deviations from the mean are the hospital's rates?
- Are the rates significantly different from previous rates?

When the data suggest further study of a particular indicator, case-by-case review occurs. Such a review focuses on the nature or cause of the variation in an effort to identify opportunities for improvement. Analysis should reveal variations in the process; unexpected patterns or trends among physicians, services, or other caregiver groups; or other factors.

Figure 8–2 is an example of Hospital XYZ's rates for a typical rate-based clinical outcome indicator, Caesarean section, which is reported to a statewide aggregate database. Hospital XYZ's rates compare favorably to the statewide averages, except for a spike in the third quarter of fiscal year 1992. This variation warrants further analysis.

Validation not only that data have been gathered correctly and concisely, but also that they represent the issue under study should be undertaken as a beginning step to explain a sudden fluctuation like the one in Figure 8–2. Following are some of the questions that should be asked:

- Was there a change in the data collection process in the third quarter of fiscal year 1992?
- Are there wide fluctuations among data collectors?
- Was the information gathered from the same source?
- Were the indicator definitions understood and applied in the same way as in previous collection periods?

- Were data collection worksheets used properly, and are all entries completed?

Once a variation has been confirmed as a genuine problem, its cause and seriousness must be determined as a major step toward resolution of the problem.

Factors that contribute to variations can often be attributed to the patient, the organization, or the practitioner. Some patient factors (e.g., severity of illness or pre-existing conditions) that may influence an outcome, causing an indicator rate to vary or even exceed a threshold, are beyond the control of caregivers. Organizational factors such as inadequate tools, poor staffing levels, ill-defined procedures, or confusing priorities may also be responsible for indicator rates. Caregivers' training, skill levels, attitudes, or preferences are other influencing forces.

The appropriate peer group must always interpret data. Professional judgment is necessary to establish standards for acceptable variations in practice. In the case of a variation in the Caesarean section rate, for example, the first step may be to request an explanation from the chairman of the obstetrics department. It is important that the evaluation process remain objective. The goal is to explain the variation, not to focus on individual practices.

Data Trending

Whether using an external comparative database or an internal database as a point of reference, decisions to analyze data should not depend on thresholds alone. There may be times when, although the threshold has

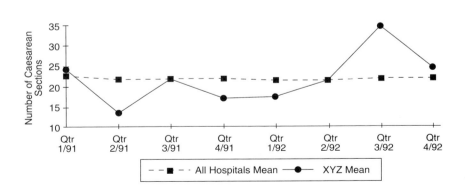

Figure 8–2 Hospital XYZ: Caesarean Section Rates

not yet been exceeded, performance has deteriorated over time. Unless the trend is checked, the threshold rate will be reached.

Data trending maximizes the organization's ability to intervene early. Analysis of both positive and negative trends is a powerful tool to influence practice patterns. Trended data are also useful to track whether an intervention is having the expected result. This is particularly useful when testing the efficiency and effectiveness of a standard of care or practice guideline.

Figure 8–3 shows length-of-stay data trended by month for the indicator pneumonia. The rate remains relatively unchanged, even though an expensive antibiotic was replaced by a less costly one in April. Practitioners are far more likely to respond to this feedback than pleas to contain costs.

Practice Guidelines

One way that hospitals have responded to public demands for value for health care dollars spent is to use practice guidelines as a means to coordinate and standardize clinical care. Practice guidelines originate in a cross-functional approach to delineate various activities and involve a time frame for the completion of these activities in the care of a particular group of patients. Practice guidelines, also called care maps or critical paths, provide the framework within which to measure the effects of certain interventions or processes on outcomes. For example, did the patient reach the milestone predicted on the predicted day? If not, why not? What, if any, was the effect of changing the sequence of care on the clinical outcome, length of stay, or cost? Variance analysis by clinical experts can be very useful in determining reasons that expected outcomes did not occur. It is also an excellent opportunity to involve front-line staff in affecting patient outcomes.

Indicators make it possible to test the efficacy of particular clinical interventions by objectively comparing the outcomes that result from each intervention. Exhibit 8–3 offers compelling evidence that the incidence of the indicator central venous line–associated bacteremia significantly decreases when a standard central venous line is replaced by an antiseptic-coated line.

Like trends, the presence of patterns of occurrences suggests that processes are not being carried out in a consistent way and that there is opportunity for improvement. Figure 8–4 graphically demonstrates that medication errors dramatically increase on weekends. Further analysis can be narrowed to the processes of administering medication that are unique to the weekends.

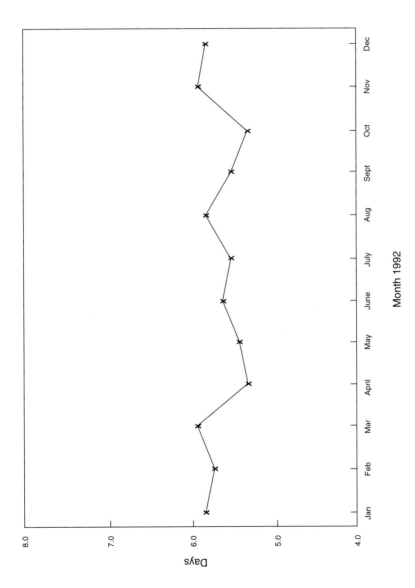

Figure 8–3 Hospital XYZ: Length of Stay for Pneumonia (Diagnosis-Related Group 89)

Exhibit 8–3 Central Venous Line (CVL)–Associated Bacteremia by Type of Line

	Standard Line	Antiseptic-Coated Line
Number of line sites	74	35
Number of cases line-associated bacteremia	15	2
Catheter Tip/Segment Positivity by Type of Line		
	Standard Line	Antiseptic-Coated Line
Number of positive tip/segments	26/60	5/30
Rate tip/segment positivity	43%	16.6%
Catheter Days and Incidence of CVL-Associated Bacteremia		
	Standard Line	Antiseptic-Coated Line
Line days	853	406
Cases of line-associated bacteremia	15	2
CVL-associated bacteremia/1,000 catheter days (incidence)	17.6	4.9

COMMUNICATION OF INDICATOR INFORMATION

Once the indicators that are appropriate and the thresholds that trigger an organizational response have been selected, responsibility for data collection and accountability for initial review should be determined. Different organizational structures address quality improvement activities in different hospitals. The dissemination of information depends on each hospital's committee and departmental structure.

Data collection and comparison to thresholds should be ongoing. Selected indicators are likely to be the purview of a particular subgroup or committee because of the need for content expertise. Initial analysis can take place at this level, or it can be the responsibility of another party (e.g., the hospital's quality improvement professional). In this latter case, data may be presented to the working group when a variance occurs and further analysis is necessary. In the absence of a variance, data may be presented monthly or quarterly, depending on the indicator. The ability of content experts to perceive patterns and trends, and to respond in a timely manner is somewhat limited by this relatively infrequent reporting, however.

The frequency of routine reviews and analyses is likely to change over time. More frequent review is necessary when the indicator is new and a reasonable threshold is still uncertain. As experience and confidence grow in the use of the indicator data, the length of time between reviews may be

Hospital Use of Clinical and Organizational Performance Indicators 141

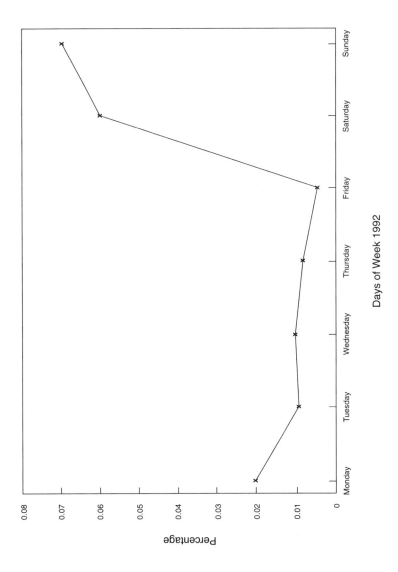

Figure 8-4 Hospital XYZ: Medication Errors

extended. Conversely, reviews may become more frequent if questions persist and changes in standards of care are anticipated.

The communication of indicator data to the facility's clinical and management leadership, quality improvement teams, or medical staff committees depends, in part, on the particular indicator, the actual performance in relation to targets, and the indicator's relationship to organizational goals and objectives. Whatever frequency and mechanism is established, it is imperative that reviews be cross-functional and multidisciplinary. The opportunity for active dialogue about the performance indicator data enhances the organizational ability to determine relationships between processes and outcomes, thereby continuously improving quality.

Because the ultimate responsibility for the quality of hospital care resides with its governing body, semiannual or annual performance indicator data are useful to keep hospital trustees informed. To be suitable for presentation to this group, quality indicator data should be concise and visually displayed, whenever possible. Comparisons between actual rates and expected rates allow board members to assess performance in an objective way. Trending data over time emphasizes trends rather than isolated events. Explanations of the data's meaning and corrective plans of action, if appropriate, should be presented.

RECOMMENDED READING

Bernstein, S.J. and L.H. Hilborne. 1993. Clinical Indicators: The Road to Quality Care? *The Joint Commission Journal on Quality Improvement* 19, no. 11:501–509.

A Compendium of Forms for Use in Monitoring and Evaluation. 1991. Oakbrook Terrace, Ill.: The Joint Commission on Accreditation of Healthcare Organizations.

Guidebook for Quality Indicator Data: A Continuous Improvement Model. 1990. Lutherville, Md.: The Maryland Hospital Association.

Hoffman, P.A. 1993. Critical Path Method: An Important Tool for Coordinating Clinical Care. *The Joint Commission Journal on Quality Improvement* 19, no. 7:235–247.

An Introduction to Quality Improvement in Health Care. 1991. Oakbrook Terrace, Ill.: The Joint Commission on Accreditation of Healthcare Organizations.

Lansky, D. 1993. The New Responsibility: Measuring and Reporting on Quality. *The Joint Commission Journal on Quality Improvement* 19, no. 12:545–551.

Leebov, W. and C.J. Ersoz. 1991. *The Health Care Manager's Guide to Continuous Quality Improvement.* Chicago: American Hospital Publishing.

Lynch, J.T. et al. 1993. The "Toward Excellence in Care" Program: A Statewide Indicator Project. *The Joint Commission Journal on Quality Improvement* 19, no. 11:519–520.

The Measurement Mandate: On the Road to Performance Improvement in Health Care. 1993. Oakbrook Terrace, Ill.: The Joint Commission on Accreditation of Healthcare Organizations.

Nadzam, D.M. et al. 1993. Data-Driven Performance Improvement in Health Care: The Joint Commission's Indicator Measurement System. *The Joint Commission Journal on Quality Improvement* 19, no. 11:492–496.

O'Leary, D.S. 1993. The Measurement Mandate: Report Card Day Is Coming. *The Joint Commission Journal on Quality Improvement* 19, no. 11:497–500.

Using Quality Improvement Tools in a Health Care Setting. 1992. Oakbrook Terrace, Ill.: Joint Commission on Accreditation of Healthcare Organizations.

Walton, M. 1986. *The Deming Management Method.* New York: Putnam Publishing Group, 3–51.

» Chapter 9 «

Collecting and Reporting of Patient Outcomes

A.J. Harper

CLEVELAND, Ohio, is a city that has pioneered many firsts in the corporate and medical community. United Way Services has its origins in Cleveland, for example. Clinically, several key advancements to patient care originated in Cleveland hospitals. The city is home to Case Western Reserve University, highly regarded for its medical and nursing education programs. In 1916, a group of Cleveland hospitals formed the first Hospital Association in the United States; this association, currently known as the Greater Cleveland Hospital Association (GCHA), today represents more than 50 hospitals in Northeast Ohio. With this history, it is not surprising that in 1988 a unique, voluntary, collaborative effort was born in Cleveland. The Cleveland Health Quality Choice Coalition is an innovative partnership between the business and medical communities formed to evaluate the delivery of health care in the Cleveland area.

In 1986, the business community initiated efforts to evaluate the delivery and cost of health care in several metropolitan areas. The objective was to develop a set of principles by which businesses could select health care services in Greater Cleveland. To that end, several site visits, presentations, and studies were conducted. Two studies of particular interest were instrumental in the movement toward the Cleveland Health Quality Choice Coalition. One study reported risk-adjusted Medicare hospital mortality data *and* Cleveland area employer claims data on hospitals in Northeast Ohio. This study reported some variation in patient outcomes between hospitals.[1] In the second study, comparison of hospital utilization rates in the Cleveland health care market with those in the Olmsted County, Minnesota health care market revealed large differences in the cost of providing care for several clinical services.[2]

The corporate leadership developed principles of market reform, which were embraced by a business community with responsibility for more than 350,000 covered lives. The market reform principles included the following:

- The key to solving cost, quality, and access problems is productivity improvement.
- The key to productivity improvement is purchaser reform.
- Purchaser reform must cease being quantity-driven and become quality-driven.

During this same time, the GCHA's executive council commissioned a task force to evaluate the current state of the art as it applies to risk-adjusted outcome measurement methodologies.[3] The task force members—physician administrators responsible for quality assurance, quality assurance directors, health information directors, and biostatisticians—began evaluating current risk-adjusted outcome measurement products established in the marketplace by reviewing published literature, conducting vendor presentations, and completing site visits at various medical centers known to be progressive in implementing risk adjustment methodology. The task force concluded that the science of risk adjustment was in its infancy and could not recommend a preferred vendor or methodology to hospitals for evaluating medical-surgical patient populations. The task force did recommend APACHE Medical Systems for evaluating care provided in intensive care units (ICUs).[4]

The corporate leadership and the GCHA began to discuss their independent findings and concluded that, with the assistance of the Cleveland Academy of Medicine of Cleveland, a coalition should be formed to determine the feasibility of identifying high-quality, cost-efficient providers of clinical services. The result was the Cleveland Health Quality Choice Coalition, whose members included:[5]

1. Academy of Medicine of Cleveland, which represents 2,900 area physicians;
2. Cleveland Tomorrow, an organization of corporate chief executive officers, whose mission is to make Cleveland a more competitive and attractive environment for business;
3. Council of Smaller Enterprises (COSE), an organization of 12,900 smaller businesses, which has recently received national attention as a model for group health care purchasing;
4. Greater Cleveland Hospital Association, an organization of over 50 hospitals in Northeastern Ohio, about half of which are located in the Cleveland Metropolitan Statistical Area; and

5. Health Action Council of Northeast Ohio, an organization of human resource and employee benefit managers from over 160 large and medium-sized corporations.

Coalition members developed two agreements, which became the foundation for the program:

1. Hospitals would adopt a standardized method to risk-adjust and report outcomes of care and patient satisfaction.
2. Businesses would use risk-adjusted outcomes of care and patient satisfaction information to encourage employees to use high-quality, cost-effective hospital services.

IMPLEMENTATION

A Systems Advisory Committee of Coalition members was commissioned to begin an evaluation and selection process. This was the beginning of the Cleveland Health Quality Choice Program. The committee identified two primary objectives:

1. to review and recommend specific indicators and patient satisfaction instruments that health care purchasers and providers can use to monitor and compare the clinical effectiveness of inpatient care provided by individual Cleveland area hospitals
2. to review and recommend a system or method to adjust for differences in the risk or "severity of illness" of patients treated by different institutions in order to compare individual hospitals fairly on the basis of the outcome indicators selected[6]

After determining that it was feasible to report patient outcomes using a risk adjustment methodology, the committee created three task forces to select a vendor to collect and report data in three areas: (1) patient satisfaction; (2) mortality, length of stay, and selected adverse events for ICU patients; and (3) mortality, length of stay, and selected adverse events for general medical, surgical, and obstetric patients. Each task force reviewed the earlier work by the business and hospital community, consulted national experts in outcome measurement, and listened to vendor presentations. The committee reached several conclusions, based on a composite of findings by the three task forces:[7]

- Outcome indicators can be used to monitor differences in performance.
- Indicators must be matched with specific diagnostic categories.

- The measurement process should start by targeting selected categories.
- Outcome must be adjusted for patient risk.
- No single risk adjustment system applies to all indicators or diagnostic categories.
- Risk adjustment should be based on patient physiological findings.
- Outcome indicators should be linked with resource consumption.
- Outcome indicators should be linked with process of care.
- The system should be flexible and responsive to the needs of purchasers and providers.

The Systems Advisory Committee emphasized the following attributes of the program:

- It was the first such effort in the United States.
- Hospitals' participation in the project was voluntary.
- All hospitals in the Cleveland area agreed to participate.
- The purpose was to evaluate services, not to assess the competence of individual physicians or to rank hospitals.
- A probable result would be a shift of patients between hospitals.

This probable shift of patients created the greatest level of interest and speculation in the hospital and physician community. Therefore, it was imperative that the selected risk adjustment methodology be not only statistically valid, but also clinically valid. In addition, benefit managers from the corporate community had to be actively involved in the measurement process; given their responsibility for negotiating and recommending managed care contracts with hospitals, they needed to understand the use and limitation of the data that would be published. Fortunately, benefit managers from the corporate and hospital community had an established collaborative relationship.

On December 27, 1991, the Quality Information Management Corporation (QIMC) was formed. Its board of trustees included representatives from the five coalition members. This board established several working committees. The Systems Advisory Committee was down-sized and eventually became known as the Quality Information Management (QIM) Committee. This technical committee addresses components of the program specific to data collection and reporting, such as audits, validation and reliability testing of models, report format, and abstract or training curriculum. A Communications Committee was formed (1) to address issues with the media and (2) to develop communication strategies to support release of the data.

METHODS OF COLLECTING AND REPORTING

Patient Satisfaction Data

In June 1990, the Patient Satisfaction Task Force recommended that the Cleveland Health Quality Choice Program contract with NCG Research, a division of Nashville Consulting Group, to implement the Hospital Quality Trends Patient Judgement System™ (PJS). The PJS survey instrument was developed through the collaborative efforts of the Hospital Corporation of America Quality Resource Group, The Rand Corporation, The New England Medical Center, Harvard Community Health Plan, and the University of California at Los Angeles (UCLA). The satisfaction instrument is used with patients who are routinely discharged from the hospital after treatment for select medical-surgical and obstetric conditions. (The task force elected to remove psychiatric and substance abuse patient populations from the survey, as task force members and vendors unanimously agreed that a radically different set of questions would be required for this population.) The purpose of the patient satisfaction survey is twofold: (1) to provide hospitals with an objective assessment of their strengths and weaknesses from the patient perspective as a basis for initiating improvements and (2) to provide purchasers with unbiased hospital comparisons to a Cleveland norm in areas important to the patient's perception of overall quality.[8]

The results of the survey are reported in 6-month intervals. Surveys are continuous throughout the year, however, with 300 patients surveyed per quarter, per hospital. Hospitals requested continuous surveying in order to take into account variations that result from such events as changes in the seasons and construction projects that present an inconvenience to the patient. More important, as Cleveland area hospitals implemented continuous quality improvement (CQI) methodologies specific to patient and customer service, the satisfaction survey instrument (renamed The Patient Viewpoint Survey™) was found to be an integral part of the data collection process (Figure 9–1).

Risk adjusting the patient satisfaction models was attempted, using patient factors such as age, sex, race, duration of time since hospital discharge, category of admission (emergent versus other), education, other hospital bills received, type of health insurance, and product line or clinical service. Of the factors examined, age and health status have the greatest influence on patient satisfaction.

The program's QIM Committee is constantly re-evaluating and revising the collection and reporting of patient satisfaction data. For example, the committee recently eliminated risk adjustment factors that have no statis-

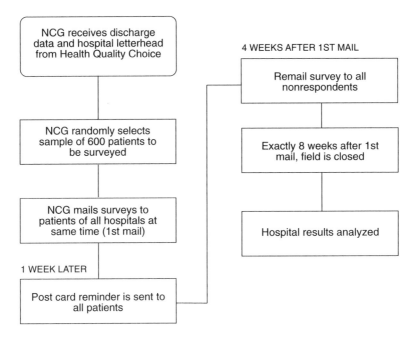

Figure 9–1 Patient Satisfaction Survey Flow Chart. Courtesy of Cleveland Health Quality Choice, Cleveland, Ohio.

tically significant impact on the patient satisfaction models. Another revision was to delete the total process patient satisfaction score in the report released to the general public. The total process score is an average of the 11 services that are surveyed (e.g., admissions, nursing care, physician care). The committee believes the total process score is misleading to the public in that the categories of daily care, information (i.e., how well a patient was informed during hospitalization), and nursing care have the highest correlation to total patient satisfaction yet are given equal weight in the total process score with billing and food service, which have the lowest correlation to total patient satisfaction (Figure 9–2). A future area of study for the QIM Committee will be the global satisfaction question (e.g., would you brag, return, or recommend?) specific to the possibility of bias toward tertiary care institutions.

Of the 1,200 surveys mailed in a year, the average response rate has fluctuated between 55 and 58 percent for medical and surgical patients. The range of response for all 29 participating hospitals is 40 to 68 percent. A separate survey of obstetrical patients is conducted once a year, and these

150   THE EPIDEMIOLOGY OF QUALITY

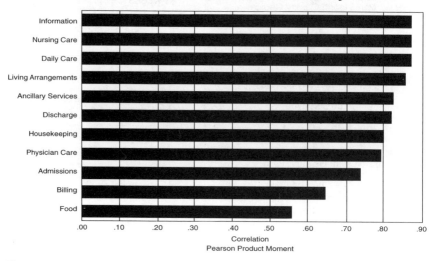

Figure 9–2 Correlation Rates in Patient Satisfaction Surveys. Courtesy of Cleveland Health Quality Choice, Cleveland, Ohio.

results are reported separately from the medical-surgical patient satisfaction survey results. In the aggregate, Cleveland hospitals have received scores in the 70th percentile, which is equivalent to a rating of "very good." Four study periods have been reported to date for medical-surgical patient populations (Figure 9–3).

Intensive Care Unit Data

Twenty-nine hospitals participating in the study operate 40 ICUs from which data are collected, and mortality and length of stay analyzed through the APACHE Medical Systems Acute Physiology and Chronic Health Evaluation III System (APACHE III). Patients who receive treatment within the four walls of an ICU make up the study population. This may include patients who are not formally admitted to the ICU, but who stay more than 4 hours for observation, and recovery room patients who stay more than 4 hours in the ICU. Excluded from the APACHE III study are:

- patients with burn injuries
- patients under 16 years of age
- patients who expire within the first hour after ICU admission

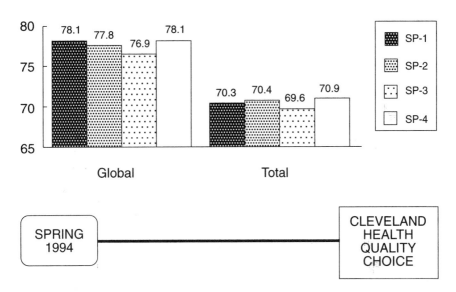

Figure 9–3 Trends in Medical-Surgical Patient Satisfaction in Cleveland. Courtesy of Cleveland Health Quality Choice, Cleveland, Ohio.

- patients who are admitted to die (within the first 4 hours after admission)
- patients admitted to ICU in full arrest who die in the first 4 hours
- patients admitted solely for routine hemodialysis or peritoneal dialysis
- observation or recovery room patients boarding in the ICU who stay less than 4 hours[9]

Risk adjustment is conducted for patient variables, including severity of condition, age, chronic health conditions, patient location prior to ICU, admission, and disease. All readmissions are excluded from the hospital's mortality analysis.

To avoid disproportionate impact from a few patients with excessively long lengths of stay, all stays in the ICU are truncated at 30 days.[10] As with mortality, adjustment for severity of condition, age, chronic health condition, treatment location prior to ICU admission, and disease are incorpo-

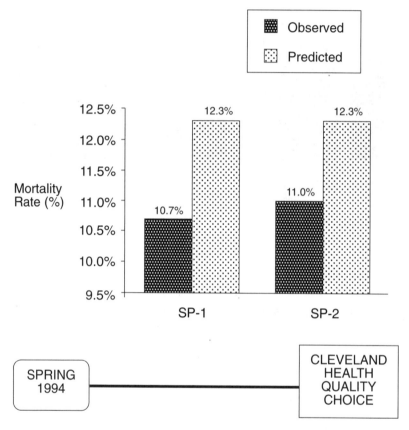

Figure 9–4 Trends in Intensive Care Outcomes in Cleveland: Mortality. Courtesy of Cleveland Health Quality Choice, Cleveland, Ohio.

---

rated into the length-of-stay models. To ensure the clinical validity of the ICU data, an expert ICU physician panel continuously evaluates the collection of data elements and explores opportunities to utilize the data for clinical study.

Trend data for Cleveland hospitals in the aggregate continue to show better than expected performance in the areas of mortality and length of stay. For the study periods reported to date, Cleveland hospitals observed a mortality rate below the predicted mortality rate (Figure 9–4). They also

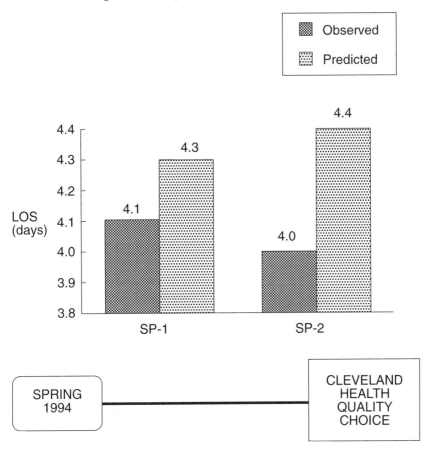

Figure 9–5 Trends in Intensive Care Outcomes in Cleveland: Length of Stay (LOS). Courtesy of Cleveland Quality Health Choice, Cleveland, Ohio.

had an observed length of stay of 4.0 days, compared to a predicted stay of 4.4 days (Figure 9–5).

Data on ICU patients are collected concurrently by the caregiving nurse in the unit. The data abstraction time ranges between 20 and 25 minutes, according to participating hospitals. There is a minimum threshold of 200 cases and 20 deaths before a hospital is eligible to report its specific data. Data are collected and reported in a wave (i.e., 3-month period). Hospitals are provided with their observed or actual outcomes against a predicted range, specific to that hospital's ICU population, displayed at a 95 percent confidence interval. For purposes of reporting, hospital performance is rated as one of the following: As predicted, Lower than predicted, or Higher than predicted. Those hospitals with outcomes lower or higher than predicted are provided with an associated p value that indicates whether their outcome is one standard deviation $p < .05$ or two standard deviations $p < .01$ outside the predicted range.

As the APACHE III database has grown in volume, the ICU physician panel has recommended reporting mortality and length-of-stay outcomes by clinical service within the ICU. Services being reported in the program include gastrointestinal, respiratory, neurological, and cardiovascular. To achieve a minimum volume threshold, data at the clinical service level are being reported over an 18-month study period. Seventy-five percent of participating hospitals must achieve a minimum volume threshold before data on that service in any hospital can be reported. Hospitals that do not achieve the minimum threshold for volume are not reported; no inference about the quality, satisfaction, or efficiency of care should be drawn about these hospitals.

Medical-Surgical Outcome Data

Because the Systems Advisory Committee could not come to agreement on a commercially available measurement system for medical-surgical outcomes, it developed a work plan to design a risk-adjusted outcome measurement system specific to the Cleveland area. Diagnosis and procedure categories were selected on the basis of their high cost, high volume, and high risk, as identified by the business *and* medical community. An additional selection criterion was the availability of data in the medical record that could be used in the risk-adjusted outcome models. Mortality and length-of-stay outcome models were built for the following:

- acute myocardial infarction
- congestive heart failure
- stroke
- pneumonia
- chronic obstructive pulmonary disease
- surgical procedures, including coronary artery bypass graft, peripheral vascular repair, gastrointestinal resection, reduction of hip fracture, laminectomy, hysterectomy, and prostatectomy

Several hospital-based adverse events were incorporated, including cardiac arrest, kidney failure, infections, and perforations.

Under the direction of the Academy of Medicine of Cleveland, expert panels of physicians were convened for each of the selected adverse event, diagnosis, and procedure categories. The panel evaluated data elements to be incorporated in developing the model's coefficients. Physicians practicing in primary, secondary, and tertiary care areas shared their clinical observations of patient populations being studied, discussed the most appropriate treatment protocols, and studied current literature on the clinical category under study. Coefficients for the medical-surgical models were shared with all hospitals—because the QIMC deemed it important to eliminate any perception of a "black box" process and to seek additional clinical input from the entire medical community. The current mortality, length of stay, and adverse event models for the medical-surgical population are continually being validated by the program's technical committee, and data elements and coefficients are under continuous review by the expert physician panels.

In a collaborative effort, Cleveland area physicians, nurses, quality management, and medical record professionals designed a data abstraction form to capture patient sociodemographic information, *ICD-9-CM* diagnostic and procedural data, key laboratory and procedural data from the admission period and subsequent hospital days, vital status, discharge destination, hospital length of stay, and the occurrence of several hospital-related complications.[11] A section of the data abstraction form and data collection instructions are illustrated in Exhibit 9–1.

The medical-surgical outcomes are reported in an identical fashion to the ICU outcomes. Calculating a predicted outcome rate for each hospital service is based on patient level clinical data. Striking a Cleveland normative value makes it possible to compare the observed (actual) and the predicted (expected) outcome rates using statistically valid methods. The predicted rates for mortality and length of stay are reported at a 95 percent confidence interval.

Data collection began with patients discharged from January through May 1991. Considered "test" data, these data were not reported to the business community, but rather were used to determine the validity and reliability of the statistical models. Currently, three additional study periods have been completed, and the database has information on well over 100,000 patients. The demographic characteristics of the database are presented in Table 9–1. Obstetrical outcome data, specifically risk-adjusted Caesarean section rates, are scheduled to be reported in the fall of 1994.

Exhibit 9–1 Abdominal X-Ray, Computed Tomography (CT), or Magnetic Resonance Imaging (MRI) Scan

---

ABDOMINAL X-RAY, CT, OR MRI SCAN:

❑ 1 No abdominal X-ray (or CT or MRI Scan) documented (skip to next test)
❑ 2 Abdominal X-ray (or CT or MRI Scan) documented

Examine all abdominal X-ray, CT, and MRI scan reports for documentation of abdominal free air, and record date (mm/dd/yy) first documented.

Finding:	Date first documented
❑ 1 Presence of free air	__ __ / __ __ / __ __
❑ 2 Free air not documented	

If the medical record contains an official report of an upper gastrointestinal X-ray (upper GI series) performed at any time during the patient's hospitalization (or in the Emergency Room contiguous to the current admission), check "Abdominal X-ray documented" and continue with this section.

If no official report of an abdominal X-ray is documented, check "No abdominal X-ray documented" and skip to next test.

Findings:
- If the presence of free air in the peritoneal space (i.e., pneumoperitoneum) was definitely documented on any report, check "Presence of free air," and record the date the definite diagnosis was first documented. If no date is documented on the report, use the medical record to establish a reasonable date on which the test was probably performed and enter that date.
- If a definite diagnosis of pneumoperitoneum (i.e., free air in the peritoneal space) is not documented, check "Free air not documented."
- "Presence of free air" should be checked and a date recorded only if at least one definite description appears on an official report. Uncertain descriptions must not be used as a basis for reporting the presence of free air.
- Note: Free air in abdomen can only be found with abdominal X-ray, not by Ultrasound.

Recommended source(s) of data: Special Procedure Report(s)
                                 Radiology Report(s)

Courtesy of Cleveland Health Quality Choice, Cleveland, Ohio.

Cleveland hospitals have demonstrated improved performance in the care of medical-surgical patients, just as they have in the care of ICU patients. Observed mortality rates and length of stay are both below those predicted for the Cleveland population in the aggregate (Figures 9–6 and 9–7).

Table 9–1 Demographic Characteristics of CHQC General Medical and Surgical Patients

	Study Period I*	Study Period II*
Total Number of Patients	27,135	58,969
Mean Age (years)	65.6	66.2
Sex		
Male	48.0%	46.8%
Female	51.9	53.2
Not documented	0.1	0.0
Race		
White	79.1	78.8
Black	18.1	19.0
Other	1.2	1.0
Not documented	1.6	1.2
Primary Insurance		
Commercial	30.4	28.8
Medicare	58.3	59.3
Medicaid	5.1	5.5
County aid	1.8	1.0
Workers' Compensation	1.1	1.1
Self pay	2.4	2.5
Uninsured	0.1	0.4
Other	0.4	0.7
Not documented	0.3	0.6
Discharge Status		
Home/self care	68.2	64.2
Transferred to a short-term hospital	2.7	2.8
Transferred to a skilled nursing facility	7.2	8.1
Transferred to an intermediate facility	3.4	3.4
Transferred to other inpatient facility		
(rehabilitative, psychiatric)	2.1	2.5
Home/home health care	10.1	13.0
Left against medical advice	0.5	0.5
Expired	5.6	5.5
Not documented	0.1	0.0
Diagnostic Category		
Acute Myocardial Infarction	7.4	7.6
Congestive Heart Failure	17.5	18.6
Coronary Artery Bypass	5.9	5.3
Gastrointestinal Hemorrhage	6.1	6.5
Hysterectomy	7.7	7.2
Laminectomy	6.6	6.0
Lower Bowel Resection	3.7	3.4
Lung Resection	1.0	1.0
Obstructive Airway Disease	10.4	11.7
Pneumonia	12.6	12.8
Prostatectomy	5.0	4.2
Reduction of Fracture	4.2	4.3
Stroke	8.8	8.6
Vascular Repair or Bypass	3.0	2.9

* Study Period I: July–December, 1991
  Study Period II: July 1992–June 1993

Courtesy of Cleveland Health Quality Choice, Cleveland, Ohio.

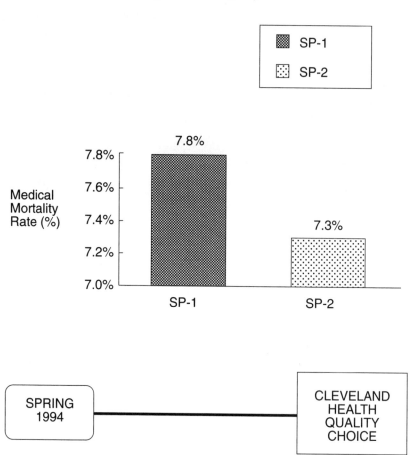

Figure 9–6 Trends in Medical Mortality Rate in Cleveland. Courtesy of Cleveland Health Quality Choice, Cleveland, Ohio.

RESPONSIBILITIES OF THE REPORT FORMAT TASK FORCE

The program's technical committee, communications committee, and a group of benefit managers formed a report format task force that was charged with (1) determining what data would be reported to whom, (2) formatting the data in easily readable tables and charts, and most important, (3) developing a training program to ensure that individuals obtaining the data were knowledgeable in their interpretation and application.

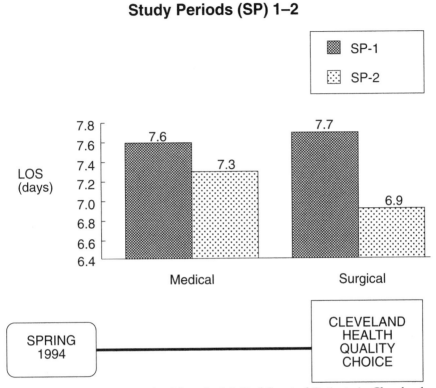

Figure 9–7 Trends in Length of Stay for Medical-Surgical Patients in Cleveland. Courtesy of Cleveland Health Quality Choice, Cleveland, Ohio.

The task force elected to produce two reports. One is an abbreviated summary report that is available to the general public and does not require any type of orientation or training. The other provides in-depth detail on program history, the methodology applied to report patient outcomes, and the actual results at a hospital-specific level. The detailed report is available to all interested parties who have completed a Cleveland Health Quality Choice Quality Measurement Training Workshop.

The training workshop is a half-day session that gives attendees an overview of the program and an opportunity to complete a case study that familiarizes them with the type of data involved in a review. Biostatistical terms, such as p value, confidence interval, and random variation, are defined. The case study then presents a scenario in which the workshop participant is a consultant to the benefit manager for a local manufacturing company that is self-insured, but utilizes a local third-party payer for processing claims, carrying out utilization review, and establishing a network of preferred providers for its employees. The benefit manager is aware of the Cleveland Health Quality Choice Program and wants to know how the third-party payer plans to incorporate the quality measurement report into the following decision-making processes:

- evaluating the existing network
- negotiating with new providers
- eliminating providers from the network
- conducting a dialogue with providers to improve quality

The workshop participants' assignment is to review a sample report of hospital outcomes and determine if the existing network is truly the best value for the manufacturing company. Participants are provided with a community fact sheet specific to the employee population, current health care expenses for the company, a fact sheet on the health care delivery systems in the community, and specific data on hospitals that participate in the program (e.g., charges, case mix index, available services, and risk-adjusted outcome data). Then, participants are asked to document and discuss issues that would affect the use of quality measurement data in managed care contract negotiations.

While developing the workshop, the members of the task force began to discuss the symbols and definitions to be used to interpret the results of data analysis. A hospital could have a result in one of five categories: performance as predicted based on the 95 percent confidence interval, or a performance better or worse than predicted at one or two standard deviations outside the 95 percent confidence interval. The task force decided to use the symbols and definitions found in Figure 9–8.

The task force also developed several caveats that describe the limitation of use and possibility for misinterpretation of the data:[12]

- Quality of care is multidimensional: Cleveland Health Quality Choice (CHQC) does not measure all dimensions of quality.
- CHQC measures a substantial proportion of patients admitted to each hospital: the analyses do not include all diagnostic categories.
- Although all comparative hospital results have been risk-adjusted using valid clinical models, it is impossible for any risk adjustment model to accurately account for all of the variation in outcomes that are due to differences in patient severity.
- Because information regarding actual hospital costs is not currently available, length of stay has been presented as a proxy measure for hospital costs and efficiency.

●	The observed performance is better than expected at the $p<.01$ level of statistical significance (i.e., there is only 1 chance in 100 that the difference between the observed and predicted performance, based on a national, normative database, arose by chance or statistical variability).
◐	The observed performance is better than expected at the $p<.05$ level (i.e., there are only 5 chances in 100 that the difference between the observed and predicted performance arose by chance).
△	The observed performance is within the normal or expected range (i.e., 95 times out of 100 the hospital's performance would be expected to fall within this range).
▬	The observed performance is worse than expected at the $p<.05$ level (i.e., there are only 5 chances in 100 that the difference between the observed and predicted performance arose by chance).
■	The observed performance is worse than expected at the $p<.01$ level (i.e., there is only 1 chance in 100 that the difference between the observed and predicted performance arose by chance).

Figure 9–8 Symbols Used To Indicate Results of Data Analysis. Courtesy of Cleveland Health Quality Choice, Cleveland, Ohio.

- The mortality and length-of-stay models and predictive equations for general medical and surgical patients should be considered "work in progress," undergoing continuous refinement.
- Although a hospital may have results that are statistically different from those predicted, it is critical to examine whether differences are of clinical or practical importance.
- The statistical analyses supporting the findings in this report do not permit a direct comparison of one hospital to another, only to the Cleveland or national mean.
- Comparative hospital results may be affected by random or chance variation.
- Comparison tables permit a degree of comparison; however, the variation noted from one point to another may be due to chance variation or difference in performance. Comparative hospital data have most value when examined serially, over time.
- There is a lag time between the study period and when data are received by hospitals. It is important to note that the lag time impacts a hospital's ability to respond to data that are reported.

Particular emphasis was placed on the tendency of the media and the public to compare or rank hospital performance. The statistical findings reported do not permit a direct comparison of one hospital to another. However, each hospital's performance can be directly compared only to the average for the comparison group. Although two individual hospitals may have different levels of performance when compared to the average, it does not necessarily follow that there is a statistically significant or clinically meaningful difference between the two individual hospitals.[13]

To bring this point home to the media and workshop attendees, the program staff presented a series of overheads that plotted actual mortality rates for all 29 hospitals in the course of a study period. The actual (observed) mortality rates were then overlaid by the predicted mortality rate for each institution based on the risk adjustment methodology incorporated by the APACHE and medical-surgical models. This series of overheads clearly illustrated that the comparison to be made is between a hospital's actual (observed) versus its predicted (expected) outcome (Figure 9–9).

Hospitals participating in the program are invited to submit commentary for inclusion in the report. These commentaries are incorporated in the report and are available to participating businesses if the following criteria are met:

1. The commentary must appear on the hospital's official (letterhead) stationery.
2. The hospital's chief executive officer must sign the commentary.
3. The QIMC must receive the commentary within 30 days of receipt of the invitation for commentary.

At the request of participating hospitals, guidelines have been adopted to clarify and assist in the correct interpretation of the report. Participants in the program, as well as other interested parties, have been instructed not to use the information for other than its intended purposes. Although several of the guidelines are repetitive of the caveats listed earlier, two of the guidelines are particularly critical and must be reiterated upon each report release:

1. Users of this information are encouraged to monitor trends and repeated measurements over time to determine the true differences in the quality of health care services among hospitals.
2. The Cleveland Health Quality Choice Program was not designed to provide marketing or advertising data for providers, purchasers, or insurers. The program's information should not be used in market-

Collecting and Reporting of Patient Outcomes 163

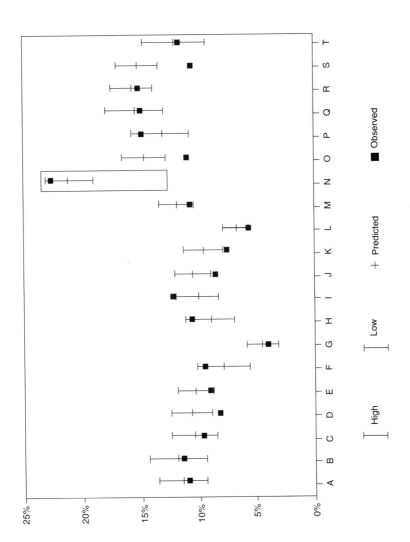

Figure 9-9 Simulated Mortality Analysis Observed (Actual) Risk-Adjusted Mortality Rate, Only—Study 3. Courtesy of Cleveland Health Quality Choice, Cleveland, Ohio.

ing or advertising in any manner that ranks or directly compares individual hospitals with one another.[14]

These guidelines represent the spirit and intent of the Cleveland Health Quality Choice Program. They are intended to make clear the mutual understanding among the partners of the coalition, and program participants have signed an agreement to abide by these guidelines.

VERIFICATION PROCESS

The QIMC has expended the majority of its resources, both time and money, on the verification process. This process focused on three distinct areas:

1. data abstraction process
2. statistical reliability of the risk adjustment methodology
3. clinical validity of the risk adjustment methodology

Data audit policies and procedures are designed to monitor hospital consistency and integrity in the abstraction process, and to ensure the quality of data received and analyzed. These policies and procedures include the following:

- All participating hospitals are required to undergo an external data review. The QIMC hired two independent auditors who rotate through the participating hospitals performing external reviews for the APACHE and medical-surgical data.
- All abstractors must be trained at a formal training session held by the vendors. A formally trained abstractor should be responsible for training new abstractors between training sessions to maintain consistency.
- A standing committee continuously evaluates and refines the data review process. This standing committee, called the Data Quality Review Board, develops minimum standards for data reliability, defines acceptable and unacceptable error rates, and facilitates the continual management of data collection. It is also available to any hospital that has problems identified by either an external or internal data review.

Initially, a 10 percent sample was taken for both APACHE and medical-surgical cases for each participating hospital. Hospitals will be expanding the sample size and reexamining the methodology for conducting external review, however. Constant monitoring is essential to avoid the possibility of systematic biasing of abstracted data. A joint venture between the

Cleveland Health Quality Choice Program and APACHE Medical Systems is expected to result in an automated data abstraction system that will have a component to facilitate the external audit process.

For the first *Hospital Data Quality Overview Report*,[15] rates of missing data higher than 5 percent were considered unacceptably high. The standard of 5 percent was chosen to separate general data abstraction problems across all hospitals from problems found in only a few hospitals, as it avoided attributing major problems to hospitals with only one missing data element out of the 20 records submitted. The 5 percent rule led to the identification of three systemwide problems that could have a profound impact on data analysis:

1. failure to record the presence or absence of specific procedures and treatments (e.g., admission chest X-ray, admission electrocardiogram, antibiotic and anticoagulant use, and performance of specific operations)
2. failure to record required findings and associated dates for diagnostic tests and procedures (e.g., magnetic resonance imaging/computed tomography scans of the brain, chest X-ray, electrocardiograms, arteriograms, pulmonary angiograms, casts, and traction)
3. failure either to record an abnormal finding or to indicate that an abnormal value was not documented in the medical record (e.g., creatinine, creatininephosphokinase (CPK), and CPK-MB values)[16]

The *Hospital Data Quality Overview Report* concluded that "low but measurable rates of missing data in numerous categories, may be regarded as a sign of a data collection process that is essentially sound but needs some 'fine tuning' to eliminate occasional sloppy data entry."[17]

The models have also been validated through statistical testing. For the mortality models, a receiver operator curve (ROC) score was calculated. For the length-of-stay models, a regression analysis methodology was applied. This methodology was applied to both the "test" data and the "live" data. The ROC scores for mortality in both APACHE and medical-surgical models were quite strong. The length-of-stay models had a wide variation, but in the aggregate also tested strongly (Table 9–2). A similar independent evaluation of the medical-surgical outcome models indicated that the models were of sound quality and worthy of continued implementation and revision.

By inviting hospital and physician commentary and releasing the medical-surgical model coefficients to the participating hospitals, the QIMC has received valuable input from clinicians to improve the models. One hospital clearly demonstrated the need to remove cystic fibrosis patients from

Table 9–2 Cleveland Health Quality Choice (CHQC): Predictive Statistics

	Test	Live
APACHE III		
Mortality (ROC)	0.89	0.90
Length of stay (R^2)	0.23	0.26
Choice		
Mortality (ROC)		
Acute MI	0.88	0.87
CHF	0.84	0.83
Pneumonia/COPD	0.90	0.86
Stroke	0.88	0.88
Length of stay	(0.16–0.46)	

ROC, receiver operator curve. R^2, regression analysis. MI, myocardial infarction. CHF, congestive heart failure. COPD, chronic obstructive pulmonary disease.

Courtesy of Cleveland Health Quality Choice, Cleveland, Ohio.

---

the pneumonia mortality and length-of-stay model, as the inclusion of these patients biased the results. Several hospitals identified a bias toward transfer patients in select medical mortality and length-of-stay models. The QIMC Board elected to continue collecting data on transfer patients, but is not currently reporting that information to the business community until additional study and testing have been completed.

The Cleveland Health Quality Choice Program is now developing a survey instrument that will be presented to purchaser and provider members of the coalition in order to begin the program evaluation process. The QIMC feels it is imperative to determine the usefulness of the data specific to quality improvement processes within the provider community and to the evaluation of managed care proposals by benefit managers.

APPLICATION OF THE DATABASE

The Cleveland Health Quality Choice data provides additional information, not previously available, that will enable employers to make better informed health care service purchasing decisions on behalf of their employees. Hospitals will use the information in their ongoing efforts to improve the quality of care that they provide their patients. Purchasers and providers have already begun to use the reported data for business and clinical purposes:

- Ætna Insurance Company is creating a subnetwork of hospitals using the Cleveland Health Quality Choice data. Ætna plans to select no more than 11 hospitals from its network, based on their quality results, and offer this network to employers at a lower cost than its entire network.[18]
- A Cleveland-based preferred provider organization is redirecting its business to hospitals that perform well in the program's quality measurement component.
- Cleveland-based Parker Hannifin Corporation created a brochure for its employees and their dependents, summarizing the performance of Cleveland area hospitals currently in the Parker Preferred network.[19]

The business community view on the performance by Cleveland area hospitals after three reports is best summarized by Patrick Casey. He said, "If the point of this Program was to identify the hospitals that do poorly, this hasn't done that. Instead, what has been identified are some providers of exceptional quality that should be a benchmark for the others."[20]

Nationally, managed care organizations have recognized the need to incorporate quality outcome measurements in their business plan. When organizations were asked to rank seven factors in order of importance to their marketplace success over the next 3 to 5 years, 69 percent ranked price first or second, followed by patient satisfaction (50 percent), and provider access (31 percent). The remaining four factors were ranked at the top as follows: quality improvement process (20 percent), national network affiliation (11 percent), clinical guide protocols (10 percent), and published outcomes (9 percent).[21] In the same study, 78 percent of the managed care organizations said that outcomes research will improve quality over the next 3 to 5 years, while only 30 percent said that it will decrease cost.[22]

Hospitals have applied all three components of the data to their continuous quality improvement processes. Patient satisfaction surveys have provided hospitals with the information necessary to identify priorities in developing customer service programs and support services for which, historically, there have been no adequate measurement instruments to establish a norm. The ability to identify the benchmark institution for a particular service or, better yet, to be the benchmark institution, is now a stated goal in many ancillary departments.

At St. John West Shore Hospital, a task force of clinicians reviewed the medical records of the 60 patients who expired in the ICU during the defined survey period. Each professional in the group examined the medical documentation and concluded that proper care had been administered in each case. The group also noted that 21 of the 60 patients, more than 33 percent, had chosen not to receive life-sustaining treatment. Thus, it was evident that St. John West Shore Hospital's ICU was functioning in some cases as a hospice, a practice that was inflating the ICU's mortality rate. Actions taken as a result of these findings have proven effective. Current studies audited through the QIMC reveal dramatic improvements. In fact, St. John West Shore's ICU mortality rate is 3.2 percent less than the national mortality rate.[23]

The QIMC began receiving requests from Cleveland-based researchers for access to the program's aggregate data. Before any analysis of data for research purposes was permitted, a Database Task Force was charged with developing guidelines to review all research projects and ensure the confidentiality of the Cleveland Health Quality Choice database. The task force developed an application for the approval of independent scientific investigation. All those requesting access to the database for research must identify (1) the major objectives of the proposed research, (2) a research plan, (3) anticipated results, (4) significance of the proposed research, (5) potential impact of the proposed research to health care providers and purchasers, and (6) agencies and/or foundations likely to fund the proposed research, when applicable.

CONCLUSION

The Cleveland Health Quality Choice Program continues to produce biannual reports on patient satisfaction and risk-adjusted outcomes for ICU and non-ICU patients. As the database grows, this information will be reported at the clinical service level. In addition, the program recently received a grant from the Cleveland Foundation to explore the feasibility of collecting and reporting functional status outcomes and indicators specific to outpatient care.

Although a formal program evaluation has not been initiated, the program has released a length-of-stay efficiency report. Combining the ICU and medical-surgical patient databases showed a reduction of more than 20,000 patient days from July, 1992 through June, 1993, correlating to a potential expense reduction of $19.6 million. The expense reduction was determined by the average cost of a patient day for hospitals participating in the program[24] (Figure 9-10).

Finally, the program has demonstrated that, at the very least, a regional effort using statistically and clinically validated risk adjustment methodologies to measure patient satisfaction and quality is feasible. Managed

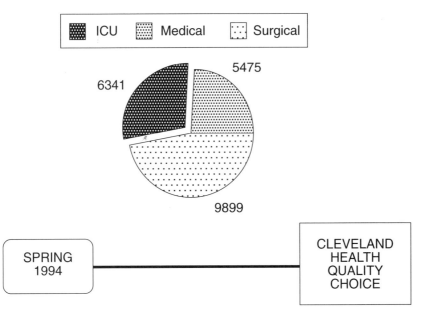

Figure 9–10 Results of the Length-of-Stay Efficiency Report. Courtesy of Cleveland Health Quality Choice, Cleveland, Ohio.

---

care organizations are only beginning to develop the information systems necessary to generate reports on the effectiveness of various medical treatments.[25] In one survey of managed care organizations, 63 percent identified data availability as one of the "barriers to effective use of outcome studies"; 54 percent identified data validity as a barrier. Among other barriers reported were cost of studies, severity adjustment, physician/hospital resistance, and confidentiality.[26] Based on its pioneering work to date, the Cleveland Health Quality Choice Program is in a position to provide managed care organizations with the ways to overcome these barriers.

NOTES

1. D.V. Shaller et al., Greater Cost Effectiveness in Health Care Delivery: Using Public and Employer Health Claims Data To Mobilize Support, *Group Practice Journal* (1992):24–35.
2. D.V. Shaller and D.J. Ballard, Olmsted County "Benchmarks," *Journal of Occupational Medicine* 33, no. 3 (1991).
3. B. Wilkenfeld et al., *Special Task Force Report on Patient Classification/Severity of Illness/Quality Assurance Systems* (Cleveland: Greater Cleveland Hospital Association, 1987).
4. APACHE Medical Systems, Inc. (AMS) is located at 1650 Tysons Blvd., Suite 300, McLean, Virginia 22101; 703-847-1400.
5. Cleveland Health Quality Choice Information Kit, Cleveland Health Quality Choice, Cleveland, Ohio.
6. D.L. Harper et al., *Report of the Systems Advisory Committee* (Cleveland: Cleveland Health Quality Choice Program, 1988).
7. Cleveland Health Quality Choice Information Kit.
8. Harper et al., *Report of the Systems Advisory Committee*.
9. D.L. Harper, *Cleveland-Area Hospital Quality Outcome Measurement and Patient Satisfaction Report* 2, no. 1 (June 1994):1.
10. Ibid., 10.
11. D.L. Harper, *Cleveland-Area Hospital Quality Outcome Measurement and Patient Satisfaction Report* 2, no. 2 (June 1994):17.
12. Cleveland Health Quality Choice Information Kit.
13. *Cleveland-Area Hospital Quality Outcome Measurement and Patient Satisfaction Report: Limitations of the Study*, Vol. 1 (Cleveland: Cleveland Health Quality Choice Program, June 1994):IX.
14. A.J. Harper et al., *Guidelines for Use of the Cleveland Health Quality Choice Information* (Cleveland: Cleveland Health Quality Choice Program, June 1993).
15. P. Hammar et al., *Cleveland Health Quality Choice Hospital Data Quality Overview Report* (Cleveland: Cleveland Health Quality Choice Program, June 1993).
16. Ibid.
17. Ibid.
18. R. Santiago, *Crane's Cleveland Business*, April 4 (1994).
19. Decisions for the Nineties; Welcome to Parker Preferred, *Parker Hannifin Corporation Report* (Cleveland: Parker Hannifin, October 1993).
20. J.M. Mazzolini, *Cleveland Plain Dealer*, June 2 (1994).
21. *Survey on Outcomes Management: Survey Report* (Princeton, N.J.: Foster-Higgins, 1994).
22. Ibid.
23. C.K. Koster and T.N. Taylor, *Clinical Study: St. John West Shore Hospital's Successful Utilization of Health Quality Choice Intensive Care Data* (Cleveland: St. John West Shore Hospital, April 23, 1993).
24. P. Lucey et al., *HealthFacts of Ohio* (Cleveland: Greater Cleveland Hospital Association, 1994).
25. *Survey on Outcomes Management*.
26. Ibid.

» Chapter 10 «

The Measurement and Management of the Quality of Ambulatory Services: A Population-Based Approach

Norbert Goldfield

BOTH private sector executives and government policy makers appreciate the dramatically increased importance of ambulatory services. In the past 10 years, the volume of ambulatory visits has increased many times,[1] and it is projected that the demand for ambulatory care will continue to rise. As a consequence, it is important that health care professionals step back and:

- Encourage policy makers, employers, and other payers of health care services to adopt key principles of quality improvement. These include measuring and managing quality on the basis of information drawn over a long period of time. This is consistent with a population-based focus in quality assessment of ambulatory services.
- Explain how quality of care measurement of populations of ambulatory services over time necessitates an understanding of the concepts of space, time, and people from both a scientific and social perspective.[2]
- Examine the many new methodologies that can be used to measure the quality of ambulatory services.
- Place the latest tools within the socioeconomic context in which they are used. Such a placement will highlight a population focus for ambulatory care quality. Thus, it can be argued that measurement of health status for low-income populations, particularly those with a chronic illness (e.g., arthritis or depression), is relevant if one takes into account community influences (e.g., level of violence, job opportunities, and educational system). In contrast, measurement of health status becomes irrelevant if one focuses on the health status of a patient with acquired immunodeficiency syndrome (AIDS) in the ab-

sence of an understanding of that patient's family, social, and community context.
- Focus their efforts on those populations most in need. Low-income and chronically ill individuals are those who will derive the greatest benefit from access to high-quality, well integrated health care services.

While the dramatic changes occurring in managed care will significantly affect the health-seeking behavior for the middle class, it is unlikely to affect low-income populations, many of whom will continue to be without insurance even after any federal legislation becomes law.

THE IMPORTANCE OF CONTINUOUS QUALITY IMPROVEMENT

Quality improvement is a process; quality of care measurement involves using tools to identify quality improvement opportunities. Measuring the quality of ambulatory services is largely a waste of time without a serious commitment to quality improvement. The initial literature on quality improvement suggests that most leaders are committed to the buzzwords, not the hard work that accompanies the implementation of the process.[3] Essential to accomplishing the hard work are:

- identification of those projects that have a significant possibility of success
- completion of the project within a brief time frame

The key commitments inherent to quality improvement, as summarized in the introduction for applicants to the Baldridge award, the U.S. government's national recognition of companies most committed to quality improvement, are:[4]

- top management involvement, including clear plans for quality leadership
- a planning process for short and long term quality improvement
- the use of data in spotting and analyzing potential problems and opportunities for improvement and a consistent data management system to ensure that accurate process information is available on a timely basis
- extensive employee involvement in the quality improvement process
- the availability of quality education and training for employees at all levels of the company

- a method for measuring customer needs and expectations, and a process for developing new or improved products that meet those requirements

Without effective implementation of these principles in conjunction with an appreciation for and implementation of Deming's Fourteen Points, it is unlikely that the quality of ambulatory services can be significantly improved. One of the greatest health care challenges will be to examine quality over time. Unfortunately, the increasingly for-profit nature of health care services does not encourage such an approach. In addition, comparison of various managed care organizations at specific points in time, rather than over the long term, is inherent to the report card concept.

THE IMPORTANCE OF A POPULATION-BASED COMMUNITY PERSPECTIVE

In a population-based approach, ambulatory services include not just the classic physician-patient encounters that occur in an office, but the whole range of services provided outside strictly inpatient hospital care. All interactions with the hospital that do not include an overnight stay constitute ambulatory services. Home-based care, whether provided by the patient only or via interactions with the medical care system, is considered ambulatory care. A population-based approach to understanding the quality of ambulatory services takes into account medications, not just those prescribed by physicians.

It is important to consider the process by which home-based services are initiated and maintained. The health care team often does more than identify the availability of support services that can be provided to a homebound patient. Increasingly, the health care team is involved in the very creation of support services so that the frail elderly patient, for example, who lives by herself can continue to stay in the community of which she has been a part all of her life.

How does one measure the quality of ambulatory care? Determining that the blood glucose level of a diabetic is normal at the time of a physician visit is an important measure of quality, for example, but the change in the measure of glucose for this same patient is even more telling if the physician is trying to identify opportunities for quality improvement. While the measure of health status of this diabetic patient at the time of visit is helpful for the physician, the change in health status over time is more important. Examining the change in health status and blood glucose levels for all diabetics over the period of a year in a managed care organization provides patients and payers with even more reliable and valid options for quality improvement. Finally, a population-based approach that is tied to an appreciation and creation of community resources provides the best means for improving the quality of care for diabetics within a community. The impact of the surrounding community on the health status of these patients is as great as that of the medical care that they are receiving. An approach that integrates an appreciation of the underlying community with a population-based perspective provides the health care organization with the most reliable and valid indication of the best ways to enhance the care of the most important customer—the patient.

CHANGES IN THE CONCEPTS OF SPACE, TIME, AND PEOPLE

A population-based perspective on ambulatory services reflects a broad understanding of the forms in which ambulatory services are delivered. Many health care professionals would state that the terms *space*, *time*, and *people* are easy to describe in relation to ambulatory care. Thus, the general population is located within a specified geographical area or ZIP code, and ambulatory care is provided within the four walls of a physician's office. Similarly, time refers to the approximately 8 hours in a health care professional's workday. Lastly, the number and type of people within a health care team (the solo practitioner versus the interdisciplinary health care team for a dying oncology patient consisting of, at a minimum, a physician, nurse-practitioner, home-based hospice team, and case manager) should be easily specified, depending on the immediate objectives of the ambulatory care administrator.

There are other approaches to examining the concepts of space, time, and people. According to the sociologist Emile Durkheim, these concepts reflect the social construction of reality as they pertain to the delivery of ambulatory care.[5] The delivery of ambulatory care is a function of:

- the needs of health care professionals. Not only is it inconvenient (and not profitable) for the vast majority of health care professionals to make home visits, but also it is physically impossible to transport the machinery of today's health care team to a more intimate setting.
- certain aspects of patient desires. For example, a patient who does not have a significant illness derives social benefits from interacting with the other patients in the office. A positive social function is served when communication occurs between patients.
- the presence of technologies that can mediate the communications process between health care professionals and patients. Telephones both enhance communication and decrease the need for visits to the

physician's office.[6] New forms of computer programming allow patients to engage actively in their own diagnostic and therapeutic process. The computer itself has had and will continue to have a great impact in radically redesigning space, time, and people needs for ambulatory services.
- the desires of health care institutions to protect their marketing position and, indeed, to increase patient belief in the need for these institutions. Despite the vast differences in the ethos of hospitals and the ethos of physician's offices, hospitals are rapidly expanding their services to include ambulatory care.
- the diagnoses that are common, associated with generous payments, or both. The recent increase in tuberculosis in many parts of the United States has resulted in increased home visits, different types of interactions between health care professionals and patients, and even hospitalizations for the sole purpose of ensuring that patients take their medication appropriately. The increased prevalence of obesity and depression is as much a function of changing social mores as it is a reflection of true clinical change. The need to medicalize depression has led to a dramatic increase in visits to health care professionals that do not require much of the treating clinician's time. The recent foray of drug companies into the managed pharmaceutical business has accentuated this problem.
- the fact that procedures have always been reimbursed at a higher rate than cognitive services. Surgeons were disproportionately represented on the Blue Cross/Blue Shield boards that determined the early patterns of reimbursement, which resulted in a tremendous discrepancy between primary care and interventions, a discrepancy that is difficult to overcome even under the resources-based relative value scale.[7]
- the importance currently attached to dying with dignity. In an era of death from chronic instead of acute illnesses, patients are rediscovering the importance of being at home and receiving services from next of kin in lieu of a hospitalization with round-the-clock nursing care. For cancer and AIDS patients, home deaths versus hospital deaths represents an excellent population-based measure of the overall ability of an integrated care system to manage the care of its patients and its ability to allocate resources in whatever setting may be appropriate.[8]

Patients and the communities of which they are a part, or ambulatory care administrators and the institutions to which they are responsible, determine many of the configurations and choices surrounding space, time, and people. For example, the organizational decision to locate a health center within a public housing project rather than to establish a one- or two-person private practice to serve a low-income population located at the edge of the community has significant implications for the health care team–patient relationship and the types of tools appropriate for measuring the quality of care delivered. Administrators of institutions must balance many competing needs in making such decisions:

- What organizational support is present?
- What financial risk is assumed in the delivery of services? While capitation introduces significant risk, it may also provide a guaranteed and known amount of cash with which several spatial arrangements can be considered.
- Are the health care professionals in different locales organizationally tied to each other? Is the administrator tied to the community? Or does the administrator consider his or her current position as a transition to the next organizational challenge?

The selection of the space for an ambulatory care setting is not merely a function of cold analytical planning of uses for available funds. It is a reflection of factors such as the social background of the administrators, the impact of community groups, and the diseases thought to be prevalent and reimbursable.

In an era of fiscal limitations, it is important to allocate space to maximize the efficiency of health care professionals. This efficiency is within the constraints of the traditional physician-patient paradigm in which an appreciation of the community has little role. Thus, the architectural design of a building serving a staff model managed care organization may consist of several examining rooms per physician in conjunction with a central, open, impersonal waiting room/reception area. This arrangement is in stark contrast with the following description of the office of a general practitioner:

> In the . . . room could be found the doctor's desk placed to get the best light from the two windows. . . . Behind the desk is the filing cabinet in which the record cards are kept; it can be reached from the chair. The couch lies along the wall. There is a space between the desk and the couch for the patient to undress, and a curtain on runners screens off this area from ceiling to floor. . . . Two big windows look out on to the garden and have gaily patterned curtains.[9]

This general practitioner's office was specifically designed with the patients in mind.

Hospitals establish ambulatory services with the primary interest of increased admissions. Recently, as patients have demonstrated less interest in attending hospital outpatient clinics, hospitals have reached out and have established satellite outpatient centers or offices for physicians. The appropriate quality of care measures may very well be different for these two different uses of space and geography. A population-based approach to quality of care measurement must include not only the four walls of the physician, but also the patient's home care. Also, the validity of a population-based perspective in designing ambulatory care spaces is heightened by today's chronic illnesses, such as AIDS, which require the active involvement of many community-based agencies.

A population-based approach to quality improvement would address the changing roles of specific professionals within the health care team and the changing nature of the team itself. For example, does a general internist have the time to use his or her moral authority to demand food stamps for a patient or cajole a job for that patient from a community-based temporary agency? Similarly, should the follow-up for a patient with panic disorder be provided by a caring primary care physician who can integrate the service with the patient's other needs, or should the follow-up be left to the clinic social worker? Is such an expenditure of time a rational use of a primary care physician's time, as these resources are currently in very short supply?

The definition of primary care services, whether provided by the physician or by the nurse, determines the ways in which health care professionals allocate their time. The efficiency rating of many managed care physicians depends on how closely they adhere to 15-minute appointments.

One managed care organization has recently calculated that, for every hour of direct visit time provided by a primary care physician for its disabled patients, one hour of nonvisit care is required.[10] Thus, it is legitimate to ask whether and how much time a primary care physician expends on case management services, for case management will be critical to the effective delivery of high-quality population-based ambulatory care. Who should be doing case management? The case manager of the future will be part of the primary care team, not a person in the administrative office of a managed care organization. Does case management time include telephone calls not only to return those made by the patient, but also to inquire about the patient's progress? Or does the managed care version of case management time consist of the efficient screening of patients who are requesting to see a specialist.

Finally, patients themselves are increasingly able to diagnose and treat their health problems. The geographical space for the diagnosis and treatment of many conditions has shifted from the physician's office to the home; the time extends beyond the 8-hour workday to reflect the emotional and geographical closeness between the patient and family or friends. With the assistance of professionals, such as nurses and attendants, patients and their families have become health care providers. The importance of the home is reflected in the distribution, by geographical setting, of dollars expended by an integrated delivery system. A managed care organization specializing in the disabled population calculates that, for its AIDS patients, $330 per member/per month is spent on home care; $160, for primary physician care; and only $47, for specialty care.[11] The quality of life of a managed care organization's chronically ill patients is improved if it can determine the best space in which the care is rendered. Over-the-counter medication has, for a long time, been a significant part of health care expenditures. Today, the increased home-based nature of care includes diagnostic testing for many conditions, including pregnancy, urinary tract infections, and AIDS. Although the treatment of AIDS requires interaction with the medical care system, many other illnesses—ranging from stomach ulcers to vaginal yeast infections—can now be treated at home. Even if the nature of the illness requires interfacing with the medical care system, such treatment modalities as the delivery of intravenous antibiotic therapy to the home for bone infections illustrate an understanding of new spaces and additional individuals involved in the care of the patient.

In summary, a community-oriented perspective toward space, time, and people might focus on the following objectives:

- understanding the health needs of the community for which one has undertaken responsibility.
- identifying community leaders and resources that could assist in improving health status. This includes community agencies and businesses interested in remaining in the community and working to improve the health status of individuals.
- emphasizing home health care, particularly those services with institutional, community-based, or family support systems in place. This entails, for example, the close coordination between case management services, hospice agencies, and the direct patient care team.
- locating practices within community spaces, such as public housing, schools, and community agencies. Such practices do not have to be staffed with full-time professionals.
- encouraging health care professionals within existing practices to open lines of communication with community agencies and leaders to enhance physician-patient communication. Although easily acted on, this activity requires significant initiative. Health care profession-

als are not reimbursed for engaging in this type of communication. This will change over the coming decade as case managers take a direct, coordinative role, rather than an administrative role.

There are projects under way that are already attempting to accomplish exactly these objectives. The most notable is The Austin Project, which articulates the following plan in the preface to its first phase report:

> This plan is to be understood as a set of recommendations for the first phase of a generation's sustained effort. The purpose of that effort will be to unite the Austin community in a determined effort to bring under control and reduce to normal proportions the now abnormal levels of teen-age pregnancy, low weight births, developmentally delayed children, family violence, school dropouts, drug use, crime and unemployment that now characterize the life of our city. This is so far as we know, the first operational urban plan organized on a systems basis; i.e., it deals explicitly with the full range of these family problems in a particular city. The Austin community has, therefore, the opportunity to demonstrate that the stagnation or degeneration that characterizes much of American urban life can be turned around.[12]

The Austin Project cannot be successful within the traditional concepts of ambulatory care space, time, and people. Ambulatory care must have a completely different definition that spreads beyond the traditional concepts of health care. The space extends to buildings, many of which are not explicitly devoted to health care services, distributed throughout the entire community. Time is not restricted to an 8-hour day. The people are not just patients and paid professionals, but people working on a pro bono basis and drawn from throughout the community.

THE TOOLS TO MEASURE AMBULATORY CARE QUALITY

Among the tools necessary for the measurement of quality are methods that can be used to classify ambulatory services, and techniques that can be used to intervene in the quality improvement process at specific points in time.

Classification Methods

Classification tools, if clinically meaningful, can provide the basis for a population-based analysis of ambulatory services. There are clinical and organizational obstacles to creating an effective ambulatory encounter system (AES), however. Ambulatory care is frequently provided for poorly understood clinical conditions.[13,14] In a large percentage of ambulatory visits, the clinician has not decided on a firm diagnosis at the completion of the ambulatory visit.[15] The poor understanding of patient motivations for seeking medical care exacerbates the poorly understood nature of many conditions seen in ambulatory services.[16] As a consequence, there is often a difference of opinion between the physician and the patient as to whether the condition was fully treated.

There are two aspects to the administrative complexity of ambulatory care. First, from a sheer geographical perspective, the patient and physician alike typically face a myriad of possible facilities from which to obtain needed services. Many of these institutions are located in widely disparate locations. Second, unlike the inpatient setting where a physician and nurse coordinate all aspects of patient care within the hospital's four walls, there is usually no such coordination in the ambulatory care arena. Even if the patient is enrolled in a health maintenance organization that aims to provide managed care, there is generally little true coordination of care. The diffuse nature of ambulatory care significantly adds to the difficulty of classifying the content of ambulatory encounters into one category.

Thus, ambulatory encounters are poorly understood for many reasons, including:

- deficiencies in clinical knowledge
- varied patient responses to the same poorly understood clinical condition
- disparate settings from which the patient can choose to obtain ambulatory services
- minimal coordination of the variegated services that many patients obtain
- facility-specific variables, such as queuing
- significant clinical discretion over numerous aspects of care, including frequency of follow-up visits and ancillary testing

The inadequacy of current coding systems further complicates the already difficult task of developing an AES. Two coding systems are in common use in the United States today: one based on the *International Classification of Disease*, 9th Edition, United States version (*ICD-9-CM*) and the other based on the *Current Procedures and Terminology—4* (*CPT-4*). CPT-4 is procedure-oriented, ICD-9-CM is oriented primarily toward inpatient care. For example, there are myriads of codes for arcane illnesses seen only in the hospital while there is little attention paid to making clinical symptoms seen on an outpatient basis (e.g., chest pain) more precise. Furthermore, few codes discriminate between different manifestations or progression of the same disease. Such differentiations are crucial to the development of a valid ambulatory classification system. Several major efforts are under way to reform currently used coding systems.[17]

To make matters worse, current *ICD-9-CM* coding rules mandate that the patient's reason for the ambulatory visit be coded as the principal diagnosis. It may be more important for the development of a reliable and valid AES that the health care professional record the code that describes the principal content, not the initial reason, of the encounter. Clearly, the coding rules and the degree to which the health care professional adheres to these rules has a significant impact on the reliability of any AES.

An AES may have three parts: (1) the classification system; (2) the "bundling" algorithm of the classes to describe the entire ambulatory encounter; and (3) part of the purpose of some systems, a payment computation. Sometimes, an AES has only a classification system. In this case, each class purports to describe ambulatory encounters that are similar from either a resource consumption or clinical cogency perspective. Other systems combine, or bundle, different classes, which together describe the ambulatory episode or visit. The bundling logic depends on clinical and nonclinical factors, such as cost and frequency of performance of a test. If the system is to be utilized for payment, the resources consumed in each class, or combination of classes, must be translated into a payment amount.

Techniques To Improve the Process

Measures that can encourage the quality improvement process include both administrative (claims-based) and clinical (medical records–based) variables. Changing the process of care emanating from these variables can lead to improvement in the process of ambulatory care. The following types of measures and actions can facilitate the quality improvement process:

1. *Surveys of a critical customer, the patient.* Not only do surveys measure patient evaluation of care, but also, wherever possible, they indicate health outcomes and the clinical quality of care. Surveys can also be obtained from other customers.
2. *Claims-based clinical quality studies.* Such studies include collection of data; provision of feedback to providers, including suggestions for improvement; and examination of new data to determine whether the process of clinical care under study has improved.

3. *Clinical indicators.* Clinical indicators for specific conditions often provide more information than do the generic screens that are still widely used today.

4. *Protocols and critical pathways.* Branching, tree-type diagrams of a clinical process are increasingly used to identify and pursue the best practices.

Although it is important to understand the technical aspects of the implementation of these tools, it cannot be overemphasized that it is the process by which the tools are implemented and developed that is critical. Satisfaction surveys are meaningless without the involvement of the audience to whom feedback will be given. Similarly, in many ways, working together with affected clinicians in developing critical pathways is more important than the pathways themselves.

Customer Satisfaction and Ratings Surveys

There is a great deal of ferment in the patient assessment arena in ambulatory care. Most health care organizations have up to now concentrated on questionnaires that elicit patient satisfaction.[18] Recently, on the theory that patient-derived data on health status are useful for comparing managed care organizations, a significant amount of research has been expended in determining the best content and context for health status questionnaires.[19] Although a great deal of research has and continues to be performed on patient-derived information, many questions remain unanswered.

Many reliable and valid patient satisfaction questionnaires are available.[20] It is advisable to retain a consultant, however, if new questions or entirely new questionnaires are deemed necessary. Even when a commercial instrument is used, many issues must be addressed in implementing a reliable and valid patient satisfaction questionnaire:

- sample frame and size
- the best format for presentation of questions
- preparation of the target audience for the questionnaire (e.g., cover letter)
- telephone versus in-person distribution methods
- administrative aspects once data have been collected (e.g., data entry, presentation of data results)[21]

More information is needed on both the development and implementation of health status questionnaires. While several questionnaires have undergone significant testing, particularly the Dartmouth Co-op Charts, numerous challenges exist with respect to their most appropriate use in the clinical setting.[22] In addition, it is necessary to decide:

- whether the patient and/or physician should complete the health status questionnaire at the
    1. initial visit and every 6 months
    2. every visit
    3. every visit only if illness is chronic
    4. at the time of enrollment into new managed care organizations
    5. on assignment with new physician
    6. at the onset of new illness
- whether the physician and/or patient should receive feedback
    1. at the time of completion
    2. on aggregated and regular basis by illness or patient

Before deciding which health status instruments to use, the user must always ask, What is the purpose of the survey? Many researchers and practitioners believe that the Dartmouth Co-op Charts are an efficient mechanism for the purpose of detecting health and social problems of large groups of patients. They have been extensively tested in adolescents and adults.[23,24] Figure 10–1 provides a sample of a Dartmouth Co-op Chart. The question of which instrument to use largely revolves around whether brief measures are as useful as longer measures in detecting and monitoring clinical change in individuals. According to Wasson and colleagues, who conducted a recent study on Co-op charts:

> Based on our experience with similar picture-and-word charts for adults, brief measures usually are not as precise as more extensive instruments. Therefore, the charts would be most appropriate for comparing treatment outcomes in a large number of subjects. These charts should not be used to monitor small changes in individuals.[25]

Another flash point for debate pertaining to health status measurement is the impact that the community of which the patient is a part has on that patient's assessment of his or her health status. For example, job availability, housing, and human services may easily affect an individual's assessment of aspects of his or her health status.

As already pointed out, questionnaires such as the Dartmouth Co-op charts are excellent measures for identifying and monitoring treatment outcomes for large groups of patients. This may be music to the ears of

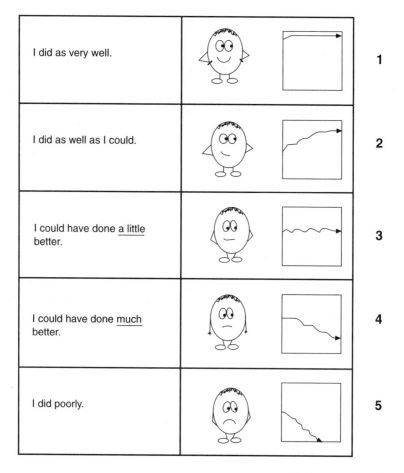

Figure 10-1 Sample of a Dartmouth Co-op Chart. *Source:* Copyright © 1990, of Dartmouth College/Co-op Project.

health care reformers who insist that it is best to evaluate managed care organizations on the basis of treatment outcomes. The process of quality improvement is enhanced in two ways when health status assessment is tied to a community focus instead of a managed care organization, however:

1. Unlike the managed competition report card concept in which managed care organizations are compared at one point in time, community oriented health status management (COHSM) emphasizes the long term and the measurement of health status over many points in time.
2. Both quality improvement programs and COHSM emphasize the need to understand and resolve systemic (i.e., including the community) problems. Improvements in job prospects and housing are integral to a community focus for health status assessment, but are very far removed from the parameters on which a managed care organization would like to be judged.

By its very nature, COHSM will have widely varying degrees of success. Many of the factors on which its success depends are embedded within the community and are resistant to the short-term interventions intrinsic to the traditional physician-patient relationship. Thus, the quality improvement process is not consistent with the philosophy of managed competition. Developing a quality improvement process surrounding the health status assessment of AIDS patients necessitates a societal perspective if one is to, for example, make an impact on the sources of AIDS transmission.

COHSM is very much in keeping with the approach espoused by the quality improvement movement in which "systemic" causes are sought and an effort is made to effect slow, incremental change. Systemic causes are important in efforts to resolve societally based issues, but they are frustrating to managed care organizations accustomed to easy decreases in rapidly rising health care costs for the worried well within the middle class. One must expect slow incremental change at best when attempting to change systemic causes endemic to low-income populations.

Claims-Based Data Studies

Several types of clinically useful information can be obtained from claims data. For example, studies of these data may show variations in care that are out of statistical control; these are typically small area variation analysis (SAVA) studies. Claims-based data are also useful for clinical quality of care studies, although these studies often necessitate referring to the medical record as well.

SAVA is an excellent screening device for determining variation in practice patterns.[26] Such studies can be performed within the plan phase of the plan-do-check-act (PDCA) cycle. An especially useful feature of SAVA studies is that they marry cost and quality concerns in their ability to identify areas of both overuse and underuse.

An important claims-based tool for the measurement of quality of ambulatory care originated in the Develop and Evaluate Methods To Promote Ambulatory Quality (DEMPAQ) project.[27] This project developed screens that essentially represent clinical indicators. Table 10–1 provides an example of the claims-based indicators developed under this project. Another important claims-based tool is a list of ambulatory care-sensitive conditions, such as diabetes and asthma, that has been developed with the postulate that access to inadequate ambulatory care results in excessive and unnecessary hospital admissions.[28] This tool has been used in many states, particularly in efforts to improve primary care access to underserved patient populations.

Thus, in a statistical process control manner, claims data can track clinical quality of care in the following categories:

- complications of surgical procedures (e.g., admissions to the hospital following cataract or hernia procedures). This criterion was chosen because it is easy to track, constitutes a variable that intersects between inpatient and outpatient care, and represents a significant failure in quality. At least one state currently collects this information.
- changes in patterns of performance of significant procedures over time to detect either underuse or overuse. One could track over time determinations of creatinine, triglycerides, cholesterol, and serum potassium levels, as recommended in the monitoring of hypertension. The indicators contained in the DEMPAQ project focus primarily on this kind of information. Rates of these procedures could be tracked in different geographical areas on a population basis.
- preventive screening in ambulatory patient groups (e.g., mammography, minor gynecological procedures, adult well care, immunizations) to identify underuse. This criterion was chosen because of its emphasis on underuse, easy trackability, and documented clinical importance.
- appropriateness in the use of ancillary (e.g., laboratory, radiology) services for particular medical or significant procedure ambulatory patient groups. For example, use of expensive ancillaries such as computed tomography (CT) scans could be tracked by hospitals. This criterion was chosen for its impact on both cost and quality. The use of relatively inexpensive ancillaries, such as a plain film, was chosen as a criterion because of the frequency of its use and as an indication of the differences by, for example, geographical area in the ambulatory patient group bundling process.

Table 10-1 Claims-Based Indicators Developed under the DEMPAQ Project

	Data source	Notes
Recommended care		
Primary care office or home visit	CPT: 90000–90080 (90100–90170) 90600–90699 90750, 90760	According to the American Board of Family Practice (ABFP), visits for diabetic patients in control should be scheduled every 3–6 months.
Hemoglobin A1c test	CPT: 83036	Since the hemoglobin A1c test gives a good indication of the level of glucose control, it is recommended every 2 to 3 months or every 6 months.
Urinalysis	CPT: 81000, 81002 81005, 81099 84180, 84185 84190	Annual urinalysis is recommended to check for proteinuria. Proteinuria is an early manifestation of diabetic nephropathy. *Note:* If urinalysis is done in the physician's office, it may not be billed separately.
Triglycerides	CPT: 84478 80061–80062 80065 83705 83720	Triglycerides should be tested annually. Hypertriglyceridemia is common in diabetics.
Total cholesterol	CPT: 82465, 824570 83700–83720 80012–80019 80050, 80053 80060–80062 80065	Total cholesterol should be tested annually because diabetics are prone to have arteriosclerosis and have an increased risk of myocardial infarction.
HDL cholesterol	CPT: 83718 80061–80062 80065	HDL cholesterol should be tested annually. In diabetics, the ratio of HDL to LDL is altered, and the risk of arteriosclerotic disease is increased.
Ophthalmology visit	CPT: 92002–92019 92225–92260	A complete eye and visual examination should be performed annually. This examination is important to detect early signs of retinopathy or cataract.
	90000–90080 90600–90699 90750, 90760	These codes are acceptable if reported by an ophthalmologist. Optometry visits in an optical shop are probably not billed to Medicare.

continues

Table 10-1 continued

	Data source	Notes
Limited-use care		
Blood glucose	CPT: 82947–82952 80006–80019	While many physicians use blood glucose to monitor patients with diabetes, guidelines from ABFP and American Diabetes Association recommend hemoglobin A1c for diabetes monitoring. In some cases, blood glucose tests may be a valuable adjunct to the hemoglobin A1c test. The blood glucose test is listed in the limited-use category because it is sometimes appropriate for patients taking insulin. However, Medicare data do not indicate which drugs are used by a patient and may not indicate whether diabetes is type I or type II. Home blood glucose monitoring is not captured by these data.

Source: Reprinted from Garnick, D.W. et al., Focus on Quality: Profiling Physicians' Practice Patterns, *Journal of Ambulatory Care Management*, Vol. 17, No. 3, pp. 54–55, Aspen Publishers, Inc., © 1994.

- inappropriate use of specific sites, such as the emergency department (e.g., rates of upper respiratory infections seen in the emergency department). There are excellent structural measures of quality that can be tracked using several ambulatory patient groups.

Analysis of claims-based data represents the first step—and only the first step—in the quality improvement process. Table 10-2 provides a summary of the tools available for the measurement of different types of ambulatory services. Health care professionals within each institution should insist that this type of data only provide a quality improvement opportunity. No organization should take final action, such as signing a managed care contract or identifying "high-quality" physicians or hospitals, solely on the basis of either administrative or clinical data. Feedback should be provided to the different institutions with a request for clarification and explanation. Only in this manner can true improvement occur.

In conclusion, recent studies have clearly demonstrated the usefulness of claims-based information for quality improvement. (Appendix 10-A provides a detailed example.) Large managed care companies have significantly expanded their claims-based capabilities. Not only do they analyze individual physician performance based on this type of information, but also they use it for compensation.[29] Thus, the private sector has already

Table 10–2 Tools Available for Measurement of Ambulatory Services

Tool	Prospective Payment	Quality Improvement	Utilization Management	CA	Public Domain or Commercial
Ambulatory patient groups	X	X	X	?	Public domain
Ambulatory care groups		X	X	X	Commercial
Diagnostic episode clusters		X	X	X	Commercial
Health Chex		X	X	X	Commercial
Patterns of treatment			X		Commercial
Value Health Profile	?	?	X	?	Commercial
United Health Care QSM	?		X	X	Commercial

Note: CA = Capitation

linked payment and quality of care issues. Implementation of ambulatory prospective payment system, as currently envisioned by the Health Care Financing Administration (HCFA), could foster this process in an important portion of the Medicare program.

Clinical Indicators

The use of clinical indicators can be helpful for monitoring commonly performed ambulatory procedures such as cataract removals, bunionectomies, and upper gastrointestinal endoscopies. Performance rates of these groups can be continuously monitored on a population basis. Several studies have focused on variations in the performance rates of such procedures on a population basis.[30] A similar effort can track upper gastrointestinal endoscopies over time. The DEMPAQ project essentially consists of clinical indicators but applies the methodology to medical conditions such as diabetes.

Protocols and Critical Pathways

Protocols are wildly heralded in some circles and deeply mistrusted in others. The issue, however, is whether they are appropriately used to improve health care for the ultimate customer, the patient. Several practical considerations must be addressed in the development and implementation of protocols and critical pathways in ambulatory services. First, protocols, algorithms, and critical pathways must be distinguished from practice guidelines and parameters. Protocols share the following characteristics:

1. a diagrammatic structure with branching logic, describing all-or-nothing choices for the provision of care
2. sequential logic, relating specific findings to appropriate actions

3. clear end points, each with a recommendation for action by health care professionals

In contrast, a guideline is a discursive, nondiagrammatic protocol.

In the ambulatory quality improvement process, protocols can be valuable tools. They are helpful in examining the processes of care, comparing different patterns or approaches, facilitating communication among physicians about clinical issues, discovering the points where clinical and administrative processes interface, planning ways to improve clinical processes, and educating or transmitting information about agreed upon methods. Protocols are least helpful when they are applied piecemeal, are unrelated to other organizational quality improvement efforts, and lack the wholehearted acceptance of both the clinical and administrative support group.

The first step in protocol development is the establishment of a climate of organizational acceptance. Initiatives may come either from senior management or from health care teams directly involved, but both groups must ultimately support the process if it is to have any impact on organizational performance.

One of the best ways to obtain a broad base of support for protocol development is to create one or more focus groups that bring together staff members from various areas and disciplines who may be affected by the proposed protocols. Regardless of the method used to secure organizational commitment to protocol development, early identification of a single individual responsible for leading the process is desirable. The leader should be trusted equally by senior management and by front-line health care personnel.

The content for protocols should meet several criteria:

- It should be comprehensive.
- It should be easily implemented, both organizationally and clinically.
- Each decision node should present a clear choice.
- Sufficient text should be attached to a protocol to explain, in detail, all the resources required to adhere to the protocol.
- Information provided should be credible to practicing clinicians.
- Protocol information should be verifiable, as well as credible.

When protocol development is complete, the even more difficult task of implementation begins. For ambulatory care, implementation is particularly challenging because of:

- the diffuse nature of ambulatory services

- varying levels of insurance coverage for different parts of an ambulatory care episode (e.g., coverage for emergency department services for asthma, but not a peak flow meter)
- difficulty in communicating among the many providers of ambulatory care

Protocols can be disseminated on paper or programmed into computer software. Numerous projects have used both approaches.[31] Resource constraints, in conjunction with managerial challenges on the best means of installing protocols in ambulatory care offices, constitute significant barriers to the effective implementation of these important ambulatory care interventions.

CONCLUSION

Clinical and policy initiatives have led to significant expansion of interest in ambulatory care services. A population-based perspective of ambulatory services would identify those groups of individuals with the greatest burden of illness and attempt to direct to these groups even a small portion of the tremendous expenditures currently spent by the managed care industry. In the short term, this approach (exemplified in the dramatic work underway in Austin, Texas) will not be the norm.

Several national organizations are involved in the evaluation of the quality of ambulatory care. Their efforts are revolving largely around the changes in managed care that emphasize managed competition. The National Committee on Quality Assurance, a collaborative effort between employers and providers, is most active with the publication of the Health Employer Data Information Set. This approach includes an examination of a variety of variables, primarily clinical indicators, and requires the documentation of the process of managed care services.

For better or worse, the challenges facing the management of quality in ambulatory services are still very much tied to issues integral to health care reform. That is, it is much more profitable to select low-risk patients for a managed care network than to manage chronically ill patients medically. This is largely a reflection of the perceived financial and managerial benefit of capitation, a payment arrangement in which a per person per year payment is given to the managed care organization, hospital, or physician. Unfortunately, adequate methods are not available to risk-adjust capitation payments to take into account the complex treatment needed by a relatively small number (compared to the larger relatively well middle class) of individuals who are either low income or have a significant chronic illness. According to the American Academy of Actuaries, "In such an environment, there would be a tremendous incentive for insurers to take advantage of any gaps in the regulatory structure and attempt to create new ways to avoid the worst health risks in order to assure survival, gain profitability, or simply minimize the risk of insolvency."[32] Even if legislation was passed mandating the enrollment of all individuals who sign up for a plan, managed care organizations have every incentive today to market only to less risky populations.

For now, the management dog of capitation reimbursement and risk selection is wagging the tail (and tale) of increasingly accurate tools for measuring ambulatory care quality. It can be hoped that there will come a time when the political commitment will exist to improve the quality of ambulatory care in the communities in which people live.

NOTES

1. R.A. Averill et al., Design of a Prospective Payment Classification System for Ambulatory Care, *Health Care Financing Review* 15, no. 1 (1993):71.
2. See M. Lock and D. Gordon, eds. *Biomedicine Examined* (Boston: Kluwer Academic Publishers, 1988).
3. Interview with S. Shortell, *Quality Connection* 3, no. 2 (1994):5.
4. U.S. Department of Commerce, Technology Administration, Introduction for Applicants to the Baldrige Award (Gaithersburg, Md.: National Institute of Standards and Technology, 1994).
5. E. Durkheim, The Dualism of Human Nature and Its Social Consequences, in *Emile Durkheim on Morality and Society*, ed. R.M. Bellah (Chicago: University of Chicago Press, 1973), 149–163.
6. J.H. Wasson et al., Telephone Care As a Substitute for Routine Clinic Follow-Up, *Journal of the American Medical Association* 267 (1992):1788–1793.
7. N. Goldfield, The High Point of Efforts To Improve Access to Health Care by Documentation, *Physician Executive* 18, no. 6 (1992):17.
8. Dr. Robert Master, Community Medical Alliance, Inc. provided me with this helpful suggestion.
9. D. Armstrong, Space and Time in British General Practice, in *Biomedicine Examined*, ed. M. Lock and D. Gordon (Boston: Kluwer Academic Publishers, 1988), 209.
10. Data courtesy of Dr. Robert Master.
11. Data courtesy of Dr. Robert Master.
12. The Austin Project, *An Investment Plan for the Young: The Austin Project, The First Phase* (Austin, Tex.: L.B.J. Library, 1992), 3.
13. A.J. Barsky, III, Hidden Reasons Some Patients Visit Doctors, *Annals of Internal Medicine* 94 (1981):492–498.
14. E.C. Nelson and D.M. Berwick, The Measurement of Health Status in Clinical Practice, *Medical Care* 27, no. 3 (1989):S79–S90.

15. J. Burnum, The Worried Sick, *Annals of Internal Medicine* 88 (1978):572.
16. J. Connelly et al., Health Perceptions of Primary Care Patients and Their Influence on Health Care Utilization, *Medical Care* 27, no. 3 (1989):S99–S109.
17. For example, 3M/Health Information Systems has been working on a new procedure coding system under contract with the Health Care Financing Administration.
18. See chapter 4 in N. Goldfield, M. Pine, and P. Pine, *Measuring and Managing Health Care Quality: Procedures, Techniques, and Protocols* (Gaithersburg, Md.: Aspen Publishers, 1994).
19. N. Goldfield, The Hubris of Health Status Assessment (Unpublished manuscript).
20. Goldfield, Pine, and Pine, *Measuring and Managing Health Care Quality.*
21. Ibid.
22. C.A. McHorney et al., The Validity and Relative Precision of MOS Short and Long Form Health Status Scales and Dartmouth Co-op Charts: Results from the Medical Outcomes Study, *Medical Care* 30(1992):MS 253.
23. J.H. Wasson et al., A Short Survey for Assessing Health and Social Problems of Adolescents, *The Journal of Family Practice* 38, no. 5 (1994):489–494.
24. J.H. Wasson et al., Benefits and Obstacles of Health Status Assessment in Ambulatory Settings, *Medical Care* 30, no. 5 (1992):S42–S49.
25. Wasson, A Short Survey for Assessing Health and Social Problems of Adolescents, 493.
26. J.W. Wennberg, Dealing with Medical Practice Variations: A Proposal for Action, *Health Affairs* 3 (1984):7–32.
27. D.W. Garnick et al., Focus on Quality: Profiling Physicians' Practice Patterns, *Journal of Ambulatory Care Management* 17, no. 3 (1994):44–76.
28. L.S. Newman et al., Health Profiles in New York City Communities, *Journal of Ambulatory Care Management* 15, no. 4 (1992):63–70.
29. N. Goldfield et al., Methods of Compensating Physicians Contracting with Managed Care Organizations, *Journal of Ambulatory Care Management* 15, no. 4 (1992):81–92.
30. N.P. Roos et al., A Population-Based Approach to Monitoring Adverse Outcomes of Medical Care, *Medical Care* 33, no. 2 (1995):127–138.
31. N. Goldfield and R. Goodspeed, Computer Assistance with Information Needs in Clinical Medicine, *New York Academy of Medicine* 88, no. 4 (1988):183–190.
32. Testimony by the Risk Adjustment Work Group, American Academy of Actuaries to the Subcommittee on Ways and Means, U.S. House of Representatives, Hearings on Risk Selection and Health Plan Adjustment Issues in Health Care Reform (mimeograph), November 9, 1993.

» Appendix 10–A «

The Plan-Do-Check-Act Cycle Examining Admissions to the Hospital Immediately Following Outpatient Surgery: A Hypothetical Case Study

PLAN

Ascertain on the basis of claims data and structured input from customers which significant procedure ambulatory patient groups and which institutions have a significant likelihood of high and low admission to the hospital following performance of a significant procedure. A uniform study to identify the process-based causes of the variation (not who is at fault) can be performed on at least a regional, if not a national, basis. Such an approach may facilitate comparisons between institutions and encourage communication between peer review organizations (PROs).

DO

A brief study can be performed initially to ascertain whether the reasons for significant changes in inpatient admissions after outpatient procedures are administrative, clinical, or both. (If they are seclusive a different PDCA study needs to be performed from the following.) A clinical study of inpatient admissions can be done to examine the processes of clinical care that may be causing the high rate of inpatient admissions following outpatient surgery. The following methodologies can be used:

1. retrospective analysis, using nurse or physician reviewers to perform a general chart review.
2. retrospective analysis, using nurse or physician reviewers to perform a chart review with a screening instrument.

3. prospective analysis of inpatient admissions after an educational intervention has been performed. The analysis involves only claims-based information with data elements added for this particular study.

Wherever possible, option 3 should be chosen for both cost and quality improvement program reasons. In addition, there is a great deal of controversy surrounding the high cost and quality assurance efficacy of the first two options.

Research has demonstrated that significant changes in inpatient admissions following outpatient surgery are typically not the result of a single incompetent provider, but rather a defect in the process of outpatient surgical care. An examination of the entire process leading up to an increase in admissions to hospitals following outpatient surgery suggests that a "problematic" provider is only one element.

CHECK

In this phase, the data are examined for reliability and validity. The data are also compared to the information gathered in the first phase to determine why differences, if any, exist. Confidential comparisons are also made between regions of the country.

ACT

Several important decisions are made in this phase:

- The PRO may work with those providers whose care is out of statistical control to determine what aspect of the process is causing undesired variation.
- The PRO then may develop, together with the affected institution, an educational intervention to address the aspect of the process that is out of statistical control. Action can be taken at this stage if the process variation is due to a problematic physician providing clearly substandard clinical care.
- The PRO can assist the provider in its continuing examination to determine if the remedies improved the process.
- For those providers whose care continues to be out of statistical control, the PRO can decide either to intensify education or to initiate punitive sanctions.

- Recognition, for example, in the form of educational awards, can be given those providers whose care is within statistical control.
- Information on this study can be made available on at least a regional, if not national, basis.
- Meetings can be held to revise specifications and standards for this quality improvement study.
- Meetings can be held with all ambulatory patient group customers to identify new areas for quality improvement studies. Such an effort will demonstrate to providers and patients alike that the research and development is devoted to true quality improvement.

» Chapter 11 «

Approaches to Quality Improvement in the British National Health Service

Liam J. Donaldson and Sir Donald Irvine

TRADITIONALLY, the measurement of quality has been anathema to many who work within the British health care system. Such aversion goes hand in hand with an idealized view of a nationalized health service that is comprehensive in its coverage, largely free at the point of access, and operated by a proud and committed workforce who believe that the system is imbued with quality by its very existence. This is a vision no longer shared by the majority of policy makers, nor by a growing number of health care professionals working within the health service, nor by the more discerning patients that it serves. Clear evidence of variable access, differential outcomes, adverse and serious incidents, and the inefficient use of resources has resulted in the widespread realization and acceptance that quality must be defined and measured.

HISTORICAL BACKGROUND OF THE NATIONAL HEALTH SERVICE

The Early Years

The National Health Service (NHS) came into being in 1948. Such a service had first been proposed in 1920 in the Dawson report,[1] and the principles and broad shape were developed with a remarkable degree of political unanimity in the 1942 Beveridge report.[2] Three elements particularly influenced the general direction of thinking and, indeed, ultimately affected the structure and modus operandi of the new service when it came into being. First, unified by the 1858 Medical Act, the medical profession was developing a strong sense of national identity and projecting that identity through the medical Royal Colleges. At the same time, the professional rivalries within medicine became more marked. Second, probably as a consequence of the Industrial Revolution, a growing sense of community responsibility tempered the innate tendency to individualism that is part of the human condition and found expression in such concepts as fairness, equity, and equality. So, the seeds were sown for a population approach to health care, complementing concern for individuals. Third, in the first half of the twentieth century, various national insurance acts extended benefits to ever widening groups of the population and stabilized prepaid health care for selected groups, using the services of family practitioners.

The Beveridge group had envisaged a publicly funded national service that would furnish the population with universal access to comprehensive health care based on a wide range of decentralized health care providers. Retained as the main provider of primary medical care, the general practitioner would become the gateway to hospital care. Overall, the intention was to have a strong, centrally planned service that would promote public health.

The actual structure, when it came, incorporated some major compromises, especially with the medical profession.[3] In particular, hospital specialists—who became salaried employees of the new service—insisted on contractual arrangements that would guarantee their clinical autonomy and leave career structures under professional control. The family physicians insisted on remaining independent contractors, functioning separately from their colleagues in the hospital and in public health medicine. The public health physicians were, to all intents and purposes, marginalized because the clinical disciplines at that time did not value public health and epidemiology.

The early NHS, like many other health care systems, had no evaluative culture so that there were few attempts to define and explicitly assess the quality of care being delivered by this entirely new entity.

Two major factors dominated the NHS agenda in its first 25 years. First, money, or rather the lack of it, soon became a major preoccupation. The idea that direct taxation alone could fund health care soon proved politically unsustainable. Consequently, the same reforming Labour government that had introduced the NHS was quickly forced to introduce drug prescription charges in an attempt to contain spiraling costs and increase revenue. Second, although a command-driven service that should by design have lent itself well to a planned approach, the health service, in fact, proved remarkably unresponsive to change. Over time, it became clear that this inertia was partly due to the excessive influence of the health care professionals, particularly the physicians, who would countenance change only on their terms and partly due to the fact that the service's management was in all respects ill-equipped for the task it was expected to perform. The result was an NHS that was perceived as capable of handling acute problems, but less effective at managing chronic illness and elective surgery. Rationing by queuing soon became the norm. So, to the outside world, the NHS was presented as a national asset; at the same time, however, it was the subject of continuing—often bitter—political debate that focused on its apparently intractable weaknesses, concerns about quality, and chronic underfunding.

In Britain, the relationship between primary and secondary care is founded on the referral system, a power-sharing agreement that originated in the nineteenth century. The agreement gave general practitioners a near monopoly on primary medical care, provided that they practiced outside hospitals, while specialists had a similar near monopoly on inpatient care. Thus, the general practitioners "owned" the patients, and the specialists "owned" the hospitals. The intense professional rivalry between general practitioners and specialists that resulted from this referral system spilled over into the structure of the new NHS, as specialization increased apace and family physicians fought to retain their position in primary medical care. Indeed, the division of the early NHS into three separate administrative parts reflected the division of the medical profession into general practitioners, hospital specialists, and public health physicians. The executive councils to which the general practitioners were under contract, the hospital management committees and hospital boards that employed the specialists, and the local municipal authorities that employed the public health physicians, thus, became equally territorial, especially in the continuous battle for influence and resources.

1984: Introduction of General Management

In 1983, Prime Minister Margaret Thatcher expressed her exasperation at the continual argument over the funding of the NHS by asking Sir Roy Griffiths, then a senior executive at a national supermarket chain with a strong quality ethos, to make an independent assessment of the NHS and determine if it could be made more efficient. In contrast with the voluminous reports on reorganization that had gone before, Griffiths wrote a 24-page letter in which he sought to bring the principles of good general management to a service that had become the largest single employer in Western Europe.[4] Griffiths noticed that at any level of the service, it was virtually impossible to say who was in charge. His proposals for reorganization were founded on the principle of decentralization and local autonomy, and producing a structure of management in which the line of accountability was clear. General managers were to be appointed to run hospitals and health authorities, and these chief executive officers were to be held personally accountable for the performance of specified services. This new managerial culture was a long way removed from the past. Although some health care professionals still deem this approach unnecessary, it will almost certainly come to stand as perhaps the most important change made to the post-war NHS and is certainly crucial to the Health Reforms that soon followed.

The 1990 NHS Reforms

Intent on bringing the costs of health care under better control, improving the quality of care, and making the professions in British society more responsive to their clientele, the Thatcher government embarked on a package of measures that became known as the Health Reforms.[5] In essence, these reforms adopted the Enthoven concept of an internal market in which purchasing and providing functions are separated.[6] Subsequently, parallels were drawn with the relationship between health maintenance organizations (HMOs) and the providers of primary and hospital health care in the United States of America.

The effects of this huge shake-up are still reverberating through the system, and it is as yet too early to predict the eventual outcome. A renewed focus on quality, its definition, and its measurement is an integral part of the internal market that has been created, however.

THE NHS OF THE 1990s: A SYSTEM TO IMPROVE QUALITY

In 1995, the NHS is still in the early phase of reforms to its structure and funding. In this new system, organizations (i.e., local health authorities and general practice fundholders) purchase health care for the population from a range of hospitals and other providers of service in an internal market. Although the market provides no true profit for the state-owned organizations in this system of care, the behavior of individuals and organizations is expected to simulate those elements of a real market that increase efficiency and improve the quality of care.

Three major forces can potentially influence quality in this public health care system.[7] First, hospitals have a tendency to compete with each other on the basis of quality, to promote their services, and to implement innovative techniques. As a result, the leaders in quality should raise the standards of others by forcing them to offer a similarly high standard of service. Second, purchasers of service tend to place contracts based on their experience of good and bad care. In theory, those hospitals that do not meet the purchasers' expectations can find their income diminishing and, ultimately, their viability threatened. Third, the opportunities for organizations purchasing health care for populations to specify in detail the quality of service to be provided can be a powerful vehicle for improvement.

The NHS reforms of 1990 created a system of public health care in which quality (not just cost) was the major driving force. The definition and assessment of quality became an integral part not only of assessing the organizations that make up the NHS, but also of judging the success of the system itself.

QUALITY IMPROVEMENT: METHODS AND INFLUENCES

Purchasing for Quality: A Population Perspective

Within the NHS, local health authorities and general practitioners continue to have responsibility for the health care needs of populations. The process of purchasing health care for these populations begins with the explicit identification of their needs and expectations (Figure 11–1). For example, a presymptomatic screening service may be required as part of a preventive strategy to reduce mortality from breast cancer, or orthopaedic operative interventions (e.g., prostheses) may be necessary for osteoarthrosis of the hip and knee joints. Next, the organization that is purchasing comprehensive care for a defined population within a cash-limited budget must make choices about which services can be funded. Then it can specify and secure the required services at the right price and of a given quality.

The recognition of purchasing as a process that can be made explicit and then improved is very much in keeping with the philosophy of total quality management. Each step has the potential to be improved (Exhibit 11–1). In establishing the requirements of the population for health care, for example, the seeking of high quality would undoubtedly involve the collection of sound epidemiological data and the valid, innovative use of those data to identify population health problems. The appropriate gathering of consumer views, wishes, and expectations would also take place at this stage. Similarly, it would be important to make sure that the research and development function contributes reliable and up-to-date information on new and emerging technologies, as well as on the clinical effectiveness of existing diagnostic and treatment techniques. There is as

Figure 11–1 Process of Purchasing Health Care. *Source:* Reprinted from L.J. Donaldson, Building Quality into Contracting and Purchasing, *Quality in Health Care*, Vol. 3, Supplement S37–S40, with permission of BMJ Publishing Group, © 1994.

---

yet little evidence of any systematic attempts to address the quality of purchasing within the NHS, though the potential is undoubtedly there.

In the whole process of purchasing health care in the NHS, much management time has been taken up with placing the contracts per se and selecting the appropriate format for quality specification. The use of explicit descriptions (or specifications) of requirements for quality as part of the contracting process, or within contracts themselves, has been seen as a key mechanism to improve quality in the post-reform NHS. A review of current practice within the NHS in the North of England has shown major weaknesses in the approaches used to improve quality through the contracting process, however (Table 11–1). Purchasing organizations were rarely formulating quality specifications that were measurable and easy to operationalize. Thus, hospitals and other providers of service were left in the position of having to interpret quality clauses, a situation unlikely to lead to the quality changes required. There was little evidence of a clear philosophy or strategy for quality improvement on the part of purchasing organizations, nor was there evidence of any systematic attempt to address established prerequisites for quality management, such as top level commitment, a customer focus, development and involvement of the workforce, and the operation of systems based on objective measures. Fur-

Exhibit 11–1 Better Purchasing: Improving the Process

Establishing Requirements	Agree Specification
• Epidemiological assessments • Consumer views • Benchmarks of performance • Implementing technological advances	• Quality criteria • Common quality values • Consumer involvement • Commitment of professional staff

Funding Decisions	Place Contracts
• Consistency with strategy • Benefits to health • Public accountability • Equity and ethics	• Choice of supplier • Detailed negotiation • Format of contract • Establish flow of information

Identify Suppliers	Quality Management
• Who could (not who should) supply • Trade offs (e.g., local access, specialist centers) • Value for money • Organization's quality commitment	• Leadership • Partnership with supplier • Evaluate performance • Further improve the process

Source: Reprinted from L.J. Donaldson, Building Quality into Contracting and Purchasing, *Quality in Health Care*, Vol. 3, Supplement S37–S40, with permission of BMJ Publishing Group, © 1994.

---

thermore, there was little evidence of the purchasing effort being directed toward improving the outcome of health care interventions. The focus of the purchasing organizations' quality strategies was much more developed at a general, hospitalwide level than at the clinical services or procedures level. Most worrying of all, there were few examples of ways to ensure contract delivery on the aspects of quality specified. As these deficiencies indicate, purchasers and providers are finding the whole process both difficult and time-consuming, especially in the beginning. It is early yet to reach informed judgments about the system.

Clinical Audit

Described as "the systematic, critical analysis of the quality of medical care, including the procedures used for diagnosis and treatment, the use of

Table 11-1 Percentage of General Quality Specifications Referring to Mechanism for Ensuring Accountability

	Medical or Clinical Audit	Patient's Charter	Complaints Procedures	Clinical Care	Service Delivery	Human Resources	Customer Care and Hotel Services	Safety and Statutory Regulations
Pre-agreed monitoring arrangements	50	88	63	0	38	13	13	13
Explicit measures of performance	0	100	25	0	38	13	13	38
Sanctions for nondelivery	0	0	0	0	0	0	0	0

resources and the resulting outcome and quality of life for the patient,"[8] clinical audit is about assessing the performance of clinicians against universal standards and bringing about beneficial change where appropriate. The 1990 British Health Reforms introduced the clinical audit, with the intention of making it an integral part of clinical practice for all health care professionals and another method for improving quality. Clinical audit was introduced into the contracts of hospital physicians as a condition of employment; general practitioners were strongly encouraged to take part, although without contractual obligation; and nurses and other health care professionals were asked to participate in clinical audits.

This NHS requirement or encouragement to participate in a quality improvement process came at a time when the medical profession had itself been exploring ways to establish standards, the peer appraisal of clinical performance, and the development of measures of health care outcome. Outstanding examples within the NHS include the confidential enquiries into maternal mortality,[9] the confidential enquiries into perioperative deaths,[10] and the national study of standard setting and performance in general practice.[11] In parallel with these developments, the Royal College of Nursing established its quality improvement project in order to assist nurses in setting and developing their own standards of care.[12] These and many similar professionally led initiatives have been seen as primarily educational, intended to promote optimum clinical practice through the process of professional development. For this reason, they tended to be unidisciplinary, to be influenced mainly by the health care professionals rather than by the users of health care, and inconsistently adopted (because their use depended on personal commitment).[13]

Within the past 5 years, under the impetus of the Health Reforms, clinical audit appears to have become more surely grounded. Concerted efforts are under way to ensure that clinical audit becomes multidisciplinary, that it involves the users of health care, and that, wherever possible, outcome is its basis.[14] A national program of research that focuses particularly on the development of clinical guidelines, valid and reliable measures of outcome, and effective implementation strategies, supports this policy.

Clinical Guidelines

The Institute of Medicine has defined clinical guidelines as "systematically developed statements to assist practitioner and patient decisions about appropriate health care for specific clinical circumstances."[15] Implicitly, guidelines emphasize rigorous science-based procedures for development and decision making about health care that involves both clinicians and patients.[16] In the United Kingdom, the term *clinical guideline* is beginning to be reserved for national consensus statements that describe best practices. Guidelines may be adapted by local clinicians to meet their own practicing circumstances, and these local adaptations are referred to as *clinical protocols*.[17]

The dissemination and uptake of guidelines in everyday clinical practice within the NHS represents a major challenge in quality improvement. Here, too, the evidence on adoption and compliance is sketchy, but it points to the need for local ownership, for the involvement of clinicians in choosing the guidelines that they should use,[18] and for carefully considered decisions about desirable change.[19] Overall, guidelines technology and management are new in Britain, and it is too early to say whether their use will become part of clinical practice and to what extent they will be beneficial for patients.

Audit Mechanisms and Strategies

Within the NHS, each hospital now has a Clinical Audit Committee. Comparable regional committees coordinate the activities of the individual hospital committees. In primary care, Medical Audit Advisory Groups (MAAGs) act as developmental agencies and assist in the implementation of clinical audits. Nationally, the Clinical Outcomes Group at the Central Government Department of Health is led jointly by the Chief Medical and Nursing Officers. Also, a national information center has been established on outcomes development. Furthermore, postgraduate medical education organizations are providing basic skills training in guidelines construction, as well as in data handling and analysis, to ensure as far as possible that each clinical unit has the capacity and the skills to carry out this quality improvement work with confidence and competence. In their totality, these arrangements are helping to bring about widespread implementation very quickly; it remains to be seen whether the investment is worthwhile.[20] Impact analysis is currently under way nationally to assess the issues that arise in the performance of this major program. First, there is the continuing question of determining how best to persuade clinicians to adopt guidelines, to follow these reasonably consistently, and to adjust their clinical practice where audit shows that changes are necessary and improvements are possible. Until very recently, the assumption was that education would bring about these behavioral changes. It is becoming increasingly clear, however, that the habit of following explicit clinical standards, monitoring performance, and changing when appropriate will develop only when standard setting and clinical audit are built into the culture and general managerial arrangements of a clinical unit.[21]

Second, as the internal market becomes established and contracting for health care becomes normal, tension grows between the concept of clinical audit as a professionally led, essentially educational activity and the concept of clinical audit as an instrument to be used in the setting and monitoring of physicians' contracts of employment. Logic suggests that both purposes are legitimate and desirable. The way forward lies in defining, in advance, the purposes to which audit will be put in any given circumstances. The matter is highly sensitive and open to misunderstanding and misinterpretation at almost every turn, however.

The third question surrounds the evaluation of audit itself. In the 5 years to 1994, the NHS allocated £160 million for medical audit support in the hospital and community health services, £42 million for infrastructure development in general medical practice, and nearly £18 million for audit in the nursing and therapy professions. Yet, there has been no formal evaluation of the effectiveness of audit. A major question is, therefore, whether this quality improvement enterprise truly represents value for money in terms both of the enhanced performance of health care professionals and of improved patient outcomes. This question so far remains unanswered.

Quality through Education: Promoting Competence and Performance

The General Medical Council, the licensing body that exercises its powers through the keeping of the medical registers under the provisions of the Medical Act, regulates the standards of medical practice and of basic medical education in Britain. Following undergraduate training, there is a preregistration (intern) year to complete basic education; at that stage, a physician becomes a fully registered medical practitioner. Thereafter, virtually all physicians must complete a period of specialty training, certainly if they wish to practice in the NHS. Specialist training is regulated by the medical Royal Colleges and Faculties, funded by the NHS, and carried out by the regional postgraduate organizations headed by postgraduate deans. Vocational training for general practice is regulated by the NHS Vocational Training Regulations of the NHS Acts and provided by regional advisors in general practice.

The Royal Colleges assess competence during and on completion of specialist training, mainly through their membership and fellowship examinations. For example, the membership examination of the Royal College of General Practitioners is the definitive standard for the assessment of the competence of those who have completed training in general practice. Practitioners may enter general practice without an objective test of competence, however, and at present, approximately one-third do.

The continuing education of established physicians has, by convention, been open-ended, relying heavily on each physician's sense of duty to keep up-to-date by regular participation in educational activities. There are also incentives to this end. For example, a (small) proportion of a general practitioner's NHS income is conditional on satisfactory attendance at accredited educational courses. Specialists may take formal study leave. Furthermore, their career progresses through a system of substantial incentive payments (merit awards) that depend, at least in part, on a subjective peer appraisal of clinical worth.

Looking to the Future

There is a growing realization that, if medical education is to be a systematic part of the process of quality improvement, it must equip physicians for a career in medicine in which change will be the rule rather than the exception, in which individualism will give way increasingly to teamwork, and in which explicit standards of practice will determine competence and performance. The future will require a greater emphasis on the continuing education of the established physician, as it is no longer tenable to assume that sound training for a specialty will sustain a physician throughout a lifetime of practice. Furthermore, recent, well documented evidence calls into question the adequacy of the profession's arrangements for identifying and handling poor practice and, therefore, for protecting the public from incompetence.[22] These factors have to be seen within a wider context in which health care consumers have higher expectations of their physicians, particularly of their competence and performance, and in which the NHS is evolving as a managed care organization that has the stated intention of placing a premium on quality.

The General Medical Council has issued recommendations to universities for basic medical education and for the reform of the intern year.[23,24] The main thrust of these recommendations is to ensure that students and newly qualified physicians are capable problem solvers, that they have a sure grounding in the science and clinical skills of medicine, and that they are capable of assessing and reassessing their values and attitudes in order to sustain optimum performance. It remains to be seen how effective the changes contained in these recommendations will be. Much will depend on the extent to which medical teachers serve as examples in their own clinical practice.

Specialist training is about to be put onto a more formal basis, with excessively long periods of training in some specialties shortened. Following the recent Calman Report[25] and in accord with European Community legislation, the General Medical Council is about to become the ultimate authority for specialist training of physicians in Britain, with the Royal Colleges and Faculties having delegated authority for the content, standards, and assessment of training to the point of specialist certification. It is likely that general practitioner training will also be brought within this framework. Building on the experience already gained in general practice, the specialties are now giving more attention to the role of the consultant as a trained educational supervisor and mentor.

The changes planned for continuing medical education are at a much earlier stage of gestation, but there is movement. Among general practitioners, for example, there is a growing interest in "the learning practice," in which practice teams take corporate responsibility for identifying their educational needs and for fulfilling these needs, both for individuals and through shared learning. The specialist Royal Colleges and Faculties are proposing to develop regular programs of planned activities aimed at ensuring continuing competence. Lastly, there is a growing realization that the link between education and the insights on a physician's performance revealed through clinical audit must be stronger and more effective. In particular, the link between identifying and remedying deficiencies in knowledge, skills, or attitudes must be closer.

The reexamination of continuing medical education goes hand in hand with the thorny question of recertification. Periodic recertification is the process in which the public, employers, and others receive visible evidence that physicians are maintaining their competence and that they are continuing to perform to a satisfactory standard. As the debate on recertification unfolds, it seems likely that there will be some public—and possibly some professional—demand for an assessment of knowledge and skill linked with an appraisal of performance at work.

Handling Poor Practice

In parallel with these educational approaches, the overall effect of which should be to help conscientious physicians deliver a higher quality service, the British medical profession and the government are considering new arrangements to identify and manage physicians who, by virtue of their seriously deficient competence or performance, place patients at risk. There are two basic approaches to this, namely, through the medical licensing process and through the contracts of physicians with their employers.

Under the Medical Act, the General Medical Council has secured agreement on the need for it to have new powers and new machinery to assess the performance of physicians who may be placing their patients at risk, to place conditions on the registration of these physicians, or to suspend them from practice altogether where serious deficiencies are demonstrated. The necessary legislation is expected to go through parliament in 1995. Meanwhile, the educational authorities have begun discussions with the government about the new arrangements that will be necessary for the retraining of physicians with demonstrable problems.

The General Medical Council has also recently agreed to provide much more explicit guidance to the profession on the duties of physicians to maintain their skills and professional performance, in particular on their duty to identify colleagues whom they believe may be putting patients at risk. For example, the council describes the duties of a physician as follows:[26]

- Treat every patient with consideration and courtesy.
- Respect patients' dignity and privacy.
- Listen to patients and respect their views.
- Give patients information in a way that they can understand, and respect their right to be involved in decisions about their care.
- Keep professional knowledge and skills up-to-date.
- Recognize the limits of a physician's professional competence.
- Be honest and trustworthy.
- Respect and safeguard confidential information.
- Ensure that personal beliefs do not adversely affect the care or treatment provided.
- Act promptly to protect patients from risk when there is good reason to believe that a physician may be unfit to practice.
- Avoid abuse of professional privileges.
- Work with professional colleagues in the ways that best serve patients' interests.

A failure to comply with this guidance may place a practitioner at risk of the charge of serious professional misconduct. Indeed, in 1994, the General Medical Council for the first time found a senior anesthetist guilty of serious professional misconduct for failing to act appropriately on identified incompetence in a consultant in a temporary position.[27]

At the same time, the NHS as an employer is reviewing its own arrangements for handling poor clinical performance in its medical workforce. Changes will emphasize the importance of preventive measures, as well as the need for clear and unambiguous guidance to NHS physicians not only about their duty to maintain their own standard, but also about their obligation to identify colleagues who may be incompetent. To this extent, the

General Medical Council and NHS are moving on a converging and mutually supportive path.

ASSESSING SERVICE PERFORMANCE

In the NHS, formal service performance assessment has traditionally been a relatively crude exercise that focused on hospital activity levels and comparisons of expenditures against budgets. Since the 1990 Health Reforms require more comprehensive and more effective measurement of service performance, the focus has been on three main issues: public accountability, quality improvement, and corporate governance.

The NHS is a publicly funded organization that consumes approximately £30 billion per year and is accountable to the public for the way in which it spends these funds. Public accountability dictates that the service be run in a way that meets the overall objectives set by the elected government, while providing the maximum possible value for the money. Both of these aims require service performance to be continually measured, monitored, and assessed.

The NHS should also be continually striving to improve quality and performance, an effort that should include providing all managers and staff with incentives to perform well. Neither individual nor organizational performance can be systematically improved without an accurate view of current performance levels and comparisons of those levels with targets, benchmarks, and best practices elsewhere. Thus, it is necessary to have an integrated planning and review system that links individual performance to organizational objectives and continually reviews performance at all levels.

The role of the boards appointed for health authorities and hospitals merits special consideration. Not only do the nonexecutive directors of such boards provide checks and balances within the management structure of the health service, but also they have potentially valuable skills and experiences to bring to local policy formulation and decision making. These roles can be effectively fulfilled, however, only if the nonexecutive directors are fully aware of the current levels of organizational performance and comparisons with objectives, targets, benchmarks, and best practices elsewhere.

There are a number of principles underlying the assessment of health service performance:

- Performance review should relate to the objectives of the health care organization.

- The degree of detail should depend on the level of assessment.
- Performance is multidimensional.
- A variety of analyses should be applied.
- Assessment should cover all policy imperatives.
- Indicators should enable comparative analysis.
- Assessments should be objective.
- The presentation of performance results should be clear, relevant, and easily assimilable.

Performance measurement often concentrates on that which is easily measurable rather than on the core purpose of an organization, which can obscure any focus. Alternately, a robust performance review that is centered on key objectives may help to prevent the organization from drifting into the pursuit of alternative, inappropriate aims.

Assessing Performance in the NHS

Service performance assessment within the NHS takes place at local, regional, and national levels. Local performance assessment is necessarily very detailed; the higher the level of review, however, the sparser the detail required and the greater the aggregation of data. For example, the assessment of hospital activity may focus on individual consultants within the hospital's own review, on procedures or case-mix groups in a health authority (purchaser) review, on specialties in a regional review, and on total hospital activity in a national assessment. These different focuses reflect the purpose of performance review at each level. Hospitals' own assessments are useful in operational decision making; health authority assessments are the basis for longer term decisions about purchasing for a population; and regional and national assessments provide information useful in developing policy and strategy. At each level, performance assessments also provide for public accountability.

Multidimensional Performance Assessment

Because performance within the NHS is multidimensional, it must be measured across a range of variables. At whichever level the assessment is being undertaken, spending levels, processes, outcomes, and the extent to which those outcomes meet organizational objectives must be assessed. An emphasis on finance, processes, or outcomes at the expense of the other dimensions will result in an unbalanced performance assessment. A performance measurement is meaningful only in the context of comparisons with those of other, similar organizations; with previous years' performances; and against targets, benchmarks, and objectives.

The NHS planning cycle currently operates on a corporate contracting process whereby purchasers contract with the NHS Executive of the Central Government Department of Health to achieve a number of annual objectives. Performance against these objectives is monitored throughout the year and reviewed at the year end to ensure delivery of agreed results.

In addition to assessments against targets and objectives, organizations need to know how their performance compares to that of other, similar organizations. For example, a health authority may be complacently achieving all of its objectives without knowing that others are achieving 50 percent more on the same measure. Therefore, it is essential to assess performance against that of other organizations. Figure 11–2 shows the percentage of purchased operations that were carried out as day cases for a group of health districts. Performance assessments are less useful without benchmarks; therefore, the best assessments are those that make wider comparisons with other organizations. Furthermore, because organizations differ in size and may even change in size over time, the best performance measures are indicators that use a denominator to standardize for these differences. Denominators such as numbers of residents, numbers of primary care contractors, and size of budget are used to permit comparative assessments. Figure 11–3 illustrates the use of standardized mortality ratios as a performance indicator.

Performance indicators allow a more informed assessment, and data are available to calculate thousands of indicators. Effective performance measurement requires the extraction of a small, manageable set of key indicators from this large potential group. Key indicators should be focused on the key aims and objectives of the organization, which in real terms within the NHS means focusing on the key policy imperatives:

- mortality rates
- waiting times
- activity levels
- financial control measures
- reduced dependence on long stay institutions
- day surgery levels
- shift of work from secondary to primary care

There are a number of policy imperatives within the NHS that must be addressed by all levels within the service.

Recent years have seen the introduction of composite performance indicators, in which different aspects of performance are weighted and combined into a single index. In the Efficiency Index, for example, different

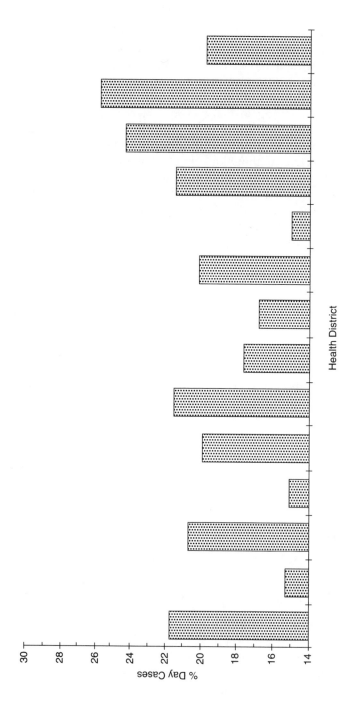

Figure 11-2 Proportion of Surgical Work Carried Out on a Day Case Basis in Health Districts of the Northern Region of England 1993–1994

Quality Improvement in the British NHS 215

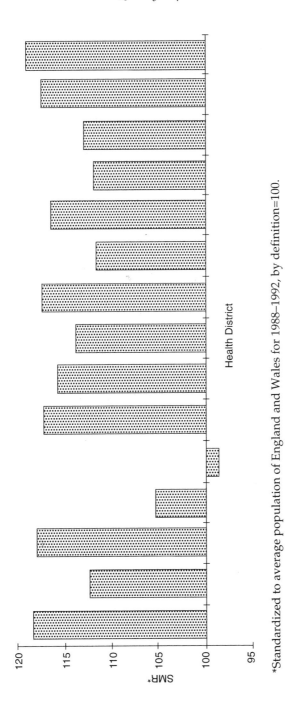

*Standardized to average population of England and Wales for 1988–1992, by definition=100.

Figure 11-3 Standardized Mortality Ratios (SMRs) for All Causes of Death in Health Districts of the Northern Region of England 1988–1992

types of activities are aggregated and weighted by cost; then year-to-year changes in resources are compared to year-to-year changes in weighted activity to produce an index. A positive index, indicating that the percentage increase in activity has been higher than the percentage increase in resources, implies that there has been an efficiency gain. The numerous conceptual and methodological problems with this index highlight some of the dangers that arise in trying to simplify the complex work of the NHS into one index.[28] If composite indicators are used in the full knowledge of their shortcomings and in conjunction with other key performance indicators rather than in isolation, however, they can help to provide a more comprehensive overview of an organization.

A prime underlying principle of performance assessment must be objectivity. Where available and relevant, hard measures should be used. In reality, the NHS is a very complex organization, and much of what it does cannot yet be measured objectively. Some softer, more descriptive assessments are necessary for some aspects of performance, such as measures of collaboration with other organizations or public consultation.

However good the performance analyses, they will not be used to maximum benefit in decision-making and review processes if they are not clearly presented to the appropriate people. Complex textual or numerical presentations can confuse and alienate managers, physicians, and the public. So, too, can particularly large or visually unattractive reports. Performance reports should be concise and well structured, and they should make full use of graphic and tabular presentations to convey the key messages to their best effect. Figure 11–4 is a graphic presentation of purchasing organization activity data, which, when accompanied by a brief explanatory text, can communicate key messages very effectively and efficiently.

Shortcomings in Current Assessments

Despite significant improvements in the quality of NHS information over the past few years, there are still weaknesses in the accuracy, relevance, and timeliness of the data produced. Objective performance assessment relies very heavily on the information that comes out of such systems; therefore, these shortcomings weaken the performance assessment. The more often information is used in the hard assessment of performance, however, the more likely managers and physicians are to strive to improve the quality of that information. As a result, the performance assessment can in itself be a positive force to improve the quality of information.

Quality Improvement in the British NHS 217

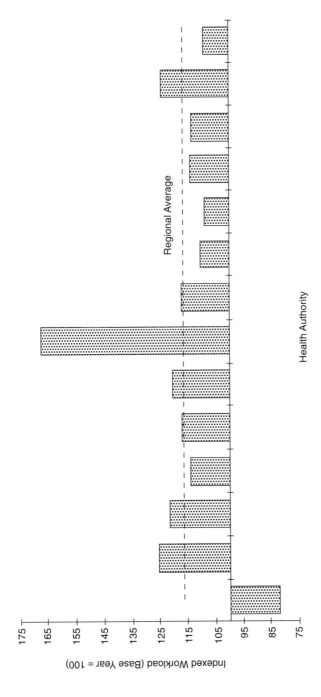

Figure 11-4 Change in Hospital Inpatient and Day Case Activity Purchased by Local Health Authorities over a 1-Year Period

Because it is impossible to get an overview of an organization from large numbers of detailed indicators and measures, the primary tools for performance measurement are often aggregated data. However, aggregation can have the effect of averaging out large discrepancies and may frequently be too blunt an instrument to address detailed issues of performance improvement. Those assessing performance must recognize the limitations of aggregated data and endeavor not to initiate inappropriate management action on the basis of broad measures.

Effective performance assessment should be comprehensive, covering all performance dimensions of all the major functions of an organization. It should also be concise, however, so that it can be easily used. These two conditions are contradictory, and it is essential to find a suitable balance between the two. Historically, NHS performance reporting has not effectively achieved this balance.

It is a truism that "you are what you measure." Health care organizations generally concentrate on those issues on which their performance is to be judged. Therefore, those assessing performance must choose measures that are related to the key issues and objectives of the organization. The selection of inappropriate measures can lead to inappropriate behavior within the organization. If managers and physicians perceive the measures being used to assess their organizations' performance as irrelevant to them, they sometimes "play the indicator game" whereby they manipulate figures to show measurements that will satisfy their assessors, without any real commitment to changing behavior in order to improve performance. This reaction is counterproductive and results in misdirected efforts.

Assessing Practitioner Performance

Competence is what a practitioner is capable of doing, and performance is what the practitioner actually does in day-to-day practice. Traditionally, competence was held to predict performance, but in fact, this is only partly true. Hence, there is increasing emphasis on the assessment of practitioner performance per se, since performance assumes competence as a basic prerequisite. In the NHS, there are currently four main approaches to the assessment of practitioner performance.

Assessment of Knowledge and Skill

The traditional approach, the assessment of knowledge and skill, is used most extensively in specialist training. The multiple choice question is still the foundation of instruments for measuring factual knowledge recall.

Knowledge, problem-solving abilities, and, to some extent, attitudes are also assessed through a variety of techniques, such as essay questions, project work, log diaries, objective structured clinical examinations, and oral examinations. The most effective overall assessments of competence use a combination of the instruments best suited to measure each attribute in question. Those combinations form the basis for the specialty examinations conducted in Britain by the medical Royal Colleges.

Observation

More common than direct observations, indirect observations may include an inspection and assessment of a physician's clinical records; interviews with the physician in his or her own office or practice; feedback from patients and colleagues with whom the physician works; and consideration of the general arrangements of a hospital unit or practice, particularly the arrangements for ensuring quality. Direct measures are used less commonly because of the sheer difficulty of collecting the evidence. Sitting in on consultations or observing a surgeon at work are the best established and most well known direct observation techniques, and they are used extensively in the apprenticeship approach to training in Britain. In general practice, video-taped consultations are also used extensively, both for teaching and for assessing consultants' skills. Problems of confidentiality have arisen with video-taped consultations, and this year the General Medical Council has introduced new guidelines to protect patient confidence.

Clinical Audit

The foremost method for assessing practitioner performance is clinical audit. When properly done, clinical audit makes it possible to compare performance against known standards and, thus, shows degrees of compliance. In the current state of development of clinical audit in Britain, it is not yet clear whether it can be or even whether it should be linked with recertification. Professional reservations notwithstanding, further developments in this direction are almost inevitable.

Direct Appraisal

British physicians have been remarkably resistant to the use of the appraisal interview, now accepted as a part of life for people in many other occupations. Formal performance appraisals are being introduced to physicians in NHS managerial appointments, however, as it is for all others in these posts. It remains to be seen whether clinicians will explore the potential of this approach both in hospitals and in general practice.

CONCLUSION

Heading for its fiftieth birthday, the NHS remains a public health care system that provides comprehensive coverage for Britain's 53 million people, largely free at the point of delivery. Although it has adhered to its founding principles, this system of care has undergone a major transformation as a result of reforms in 1990 to simulate market behavior and benefits within a nonprofit environment. The impetus for these changes was not merely cost containment; it was also a drive for a new focus on quality.

The formal assessment of quality in the NHS is a developing art that has improved considerably since the 1990 reforms placed issues of performance firmly in the spotlight. Significant constraints continue to prevent consistently sound assessment within the service, however. The task of managers, physicians, and other professional staff throughout the NHS is to inculcate a performance culture within which all organization members are continually monitoring and measuring the quality of their own work to enable them continually to improve it.

NOTES

1. Dawson, *Interim Report on the Future Provision of Medical and Allied Services*, Cmd 693 (London: HMSO, 1920).
2. W. Beveridge, *Social Insurance and Allied Services*, Cmd 6404 (London: HMSO, 1942).
3. W. Laing, *Managing the NHS Past, Present and Agenda for the Future* (London: Office of Health Economics, 1994).
4. R. Griffiths, *NHS Management Enquiry* (London: Department of Health and Social Security, 1983).
5. Secretaries of State for Health, Wales, Northern Ireland and Scotland, *Working for Patients*, Cm 555 (London: HMSO, 1989).
6. A. Enthoven, Reflections on Management of the National Health Service. (London: Nuffield Pronvincial Hospitals Trust, 1985).
7. R.J. Donaldson and L.J. Donaldson, *Essential Public Health Medicine* (London: Kluwer, 1993).
8. Secretaries of State, *Working for Patients*, 39.
9. See the reports on confidential enquiries into maternal deaths in England and Wales 1952/54, 1955/57, 1958/60, 1961/63, 1964/66, 1967/69. (London: HMSO).
10. E.A. Campling et al., *The Report of the National Confidential Enquiry into Perioperative Deaths* (London: Royal College of Surgeons of England, 1989).
11. D.H. Irvine et al., Performance Review in General Practice: Educational Development and Evaluative Research in the Northern Region, in *In Pursuit of Quality?* ed. D.A. Pendleton et al. (London: Royal College of General Practitioners, 1986).
12. Royal College of Nursing, *Quality Patient Care: The Dynamic Standard Setting System* (London: 1990).

13. Audit Commission for Local Authorities and the National Health Service in England and Wales, *Minding the Quality: A Consultation Document on the Role of the Audit Commission in Quality Assurance and Health Care* (London: 1992).
14. NHS Management Executive, *The Evolution of Clinical Audit* (London: Department of Health, 1994).
15. Institute of Medicine, Committee on Clinical Practice Guidelines, *Guidelines for Clinical Practice* (Washington, D.C.: National Academy Press, 1992), 2.
16. K.N. Lohr, Guidelines for Clinical Practice: Applications for Primary Care, *International Journal for Quality in Health Care* 6 (1994):17–25.
17. Royal College of General Practitioners, *Quality and Audit in General Practice: Meanings and Definitions* (London: 1994).
18. J. Grimshaw and I.T. Russell, Achieving Health Gain through Clinical Guidelines: I. Developing Scientifically Valid Guidelines, *Quality in Health Care* 2 (1993):243–248.
19. J. Newton et al., Education Potential of Medical Audit: Observations from a Study of Small Group Setting Standards, *Quality in Health Care* 1 (1992):256–259.
20. National Audit Office, *Auditing Clinical Care in Scotland* (London: HMSO, 1994).
21. D.H. Irvine and L.J. Donaldson, Quality and Standards in Health Care, *Proceedings of the Royal Society of Edinburgh* 101B (1993):1–30.
22. L.J. Donaldson, A Study of Doctors with Problems in an NHS Workforce, *British Medical Journal* 308 (1994):1277–1282.
23. General Medical Council, *Tomorrow's Doctors* (London: 1993).
24. General Medical Council, *Recommendations on General Clinical Training* (London: 1992).
25. Working Group on Specialist Medical Training, *Hospital Doctors: Training for the Future*, Chairman Dr. Kenneth Calman (London: Department of Health, 1993).
26. General Medical Council, *Good Medical Practice* (London: 1994).
27. C. Dyer, Consultant Found Guilty of Failing To Act on Colleague, *British Medical Journal* 308 (1994):809.
28. L.J. Donaldson et al., Lanterns in the Jungle: Is the NHS Driven by the Wrong Kind of Efficiency? *Public Health* 108 (1994):3–9.

» Chapter 12 «

The Public Health Paradigm

C. Patrick Chaulk

POPULATION-BASED health data play a unique role in the development of effective public health policy. The public health paradigm is based on population-oriented activities. It is community-focused (often community-based); addresses populations rather than individuals; and frequently emphasizes prevention, screening, and early disease intervention within a population. Water fluoridation and restaurant inspection services are traditional population-based public health measures. Both are directed at populations, not at individual patients. The former represents primary prevention; the latter, secondary prevention.

One of the justifications for public health interventions is the fact that certain health care activities are more effective when conducted from a population perspective.[1] The task of water fluoridation would be cumbersome and unwieldy if performed house by house, for example. Moreover, reducing the incidence of disease in a community frequently requires interventions beyond the traditional curative services delivered to individual patients. The control of sexually transmitted disease (STD), for example, requires considerably more than treatment of the index case. Each diagnosis of STD reflects at least one, often more, additional case in the community, since the patient acquired his or her infection from someone else. Treating only the index case leaves any additional cases to act as potential reservoirs for further STD transmission. Public health strategies involve services intended to break this cycle of STD infection.[2] Services provided through public programs include:

- employing epidemiological techniques to identify individuals at greatest risk of STD infection, thus increasing program cost-effectiveness

- compiling public health data sets to define the natural history and changing epidemiology of STDs over time
- testing for difficult to diagnose syndromes, such as those associated with syphilis and chancroid
- screening for co-infections, a phenomenon more common in the field of venereology
- improving compliance through observed therapy, enhanced counseling, and follow-up for test of care
- conducting community surveillance to assess disease control effectiveness
- screening and offering referrals for associated problems, such as ectopic pregnancy, pelvic inflammatory disease, co-infection with human immunodeficiency virus (HIV) or tuberculosis
- providing information on family planning

Generally, teams of communicable disease specialists, including nurse-practitioners, physician assistants, public health nurses, disease control and outreach specialists, and physicians, provide these services.

The traditional medical paradigm emphasizes the individual, and the measure of its success is the resolution of illness in the individual patient. Although the traditional medical model includes screening and prevention services, its primary focus is cure; thus, a measure of its performance is changes in death and disability rates related to specific disease processes among individuals. Communicable disease control programs (e.g., mass vaccination and tuberculosis control programs) apply population approaches and measure their performance in the prevalence of selected medical conditions in a predetermined population.

These two paradigms differ in another aspect relevant to data. Under the medical model, the individual undergoes comparatively little change over time, with the notable exception of change brought about by the aging process or by external events such as injury. In contrast, the composition of a population nearly always changes over time through immigration, births, or deaths. These events sometimes produce profound changes in disease patterns relevant to public health programs. For example, immigration can dramatically alter the incidence of communicable diseases in a given region. Data that reflect these time trends are particularly important for public health decision making, and the epidemiological concepts that underlie public health policy in a community are important in the assessment of population-based needs.

In the real world of public policy, programmatic decisions and resource allocation frequently must be made without the benefit of double-blind prospective or case control studies. Policy makers and politicians must often make choices using population-based and cross-sectional data. Although comparatively imprecise, and less powerful than controlled studies, cross-sectional and population-based data may be all that are available to help make judgments about program efficiency, quality, and continuation.

The importance of public health data in the epidemiologic approach to such public health problems can be demonstrated in an examination of the problem of tuberculosis. Its etiology is well known, and its treatment and management have been clearly defined, although not necessarily adequately disseminated. More importantly, its predicted demise has been countered by a resurgence in the United States in recent years. In some inner cities, it is epidemic.

As with other public health programs, tuberculosis control involves a range of nonclinical services. These include:

- using epidemiologic strategies to track disease trends and the natural history of tuberculosis within different populations
- identifying social and biologic risk factors for infection and progression to fulminant disease
- identifying and screening individuals exposed to active cases (case contacts) and documenting their treatment with preventive therapy (to prevent progression to a stage of contagious disease)
- assessing tuberculosis control program performance by monitoring treatment compliance, e.g., the percentage of patients completing treatment
- monitoring drug susceptibility patterns of tuberculosis organisms to identify new drug resistant strains

TUBERCULOSIS AS A PUBLIC HEALTH PROBLEM

Tuberculosis retains an unusual place in the field of communicable disease. Its historical background has been extensively documented.[3,4] Today, between 1.5 and 2 billion people are infected with tuberculosis, making it the most common infectious disease in the world. In the United States, an estimated 1 million people are infected with tuberculosis; some 26,673 new cases of tuberculosis were reported in 1992.[5] Rates are high among crowded inner city populations, among the poor, and, increasingly, among young adults.[6] Tuberculosis may infect any body organ, including the blood and brain, and its protean ability to develop resistance to the most potent antimicrobials make its management especially challenging.

The etiology of tuberculosis, *Mycobacterium tuberculosis*, was identified more than 100 years ago by Robert Koch.[7] Available medical technology permits presumptive identification of the microorganism. The use of chemical staining techniques provides rapid evidence of *M. tuberculosis* in body samples, typically sputum.[8] It does not, however, reveal the viability of the microorganisms. Accurate identification can be made in 2 to 3 weeks by employing more advanced technological tests.[9] Antimicrobial sensitivity cannot be established, however, until some 6 to 8 weeks after specimen collection and culture.[10] Thus, although cumbersome, the process of identification leads to a definitive diagnosis in most instances.

The public's perceptions notwithstanding, tuberculosis is not highly contagious.[11] Most individuals who become infected (as many as 90 percent) are able to contain the infection in a dormant state (latent tuberculosis), and thus do not develop clinical disease. Therefore, among individuals with competent immune systems, less than 10 percent develop active disease, that is, the classic symptoms of tuberculosis: severe and productive cough, prolonged fever, night sweats, and weight loss. Among those individuals who become ill, half do so within the first 1 to 2 years after they were infected; the other half do so much later in life.[12]

New molecular-based techniques are capable of determining the specific organism responsible for epidemics, thus providing a unique role for genetics in public health disease control.[13,14] It is now possible, for example, to trace communicable disease transmission among individuals through genetic "fingerprinting" of the infecting organism. Public health specialists have used genetic analysis to demonstrate that a single individual with tuberculosis generated a mini-outbreak by infecting nearly 30 people.[15] Such findings highlight the importance of case finding and chemoprevention for case contacts.

Tuberculosis registries contain all confirmed cases reported to public health authorities.[16] Case finding through public programs and mandatory reporting of confirmed tuberculosis cases permit public health officials to determine disease trends and define populations at risk; identify the co-factors that contribute to disease acquisition; and demarcate geographical catchment areas with high infection rates. Ultimately, such monitoring becomes the basis for quality assessment of program performance.[17,18] From a public health perspective, therefore, centralized data collection makes it possible to establish disease patterns over time. Obtaining case factors such as gender, age, and other selected risk factors further enhances public disease control policy by shedding light on the natural history of the disease, the development of new risk groups for *M. tuberculosis* infection, the evolution of drug-resistant species, and new risks for acquisition and transmission. In the absence of such data, public health policy will be severely compromised.

PUBLIC DATA AND PUBLIC POLICY ON TUBERCULOSIS

Population-based data have been used to develop disease trends and identify groups at risk of infection with *M. tuberculosis*. For example, case finding and reporting suggest that infants and children under 5 years of age, adolescents, and the elderly are at particular risk of tuberculosis. Public data also reveal regional variations; rates are highest in states with large urban and immigrant populations. Immigration, itself a stable factor during the 1980s, accounts for roughly 25 percent of all tuberculosis each year. In addition, tuberculosis rates differ according to ethnicity. Nonwhites are much more likely than are whites to be infected with tuberculosis[19] (see Table 12–1).

Using tuberculosis trends, Grigg developed his classic "tuberculosis wave" to illustrate the expected impact of introducing a "new" strain of tuberculosis into a fully susceptible and closed population (i.e., one that is stable and not influenced by external factors).[20] Contrary to other infectious disease curves, which sustain disease transmission for only brief periods of time, this epidemic wave encompasses several centuries (Figure 12–1). The tuberculosis curve is characterized by a steep sloping ascending mortality curve that peaks during the first 100 years. This curve reflects the increased risk of early progression from infection to clinical illness. Stag-

Table 12–1 Reported Tuberculosis Cases by Race and Ethnicity in the United States, 1991

Race/Ethnicity	Number of Cases	Percent
Non-Hispanic White	7,709	29.3
Non-Hispanic Black	9,536	36.3
Hispanic	5,330	20.3
Asian/Pacific Islander	3,346	12.7
American Indian/Alaskan Native	345	1.3
Other/Unknown*	17	0.1
Total	26,283	100.0

*Other includes blacks and whites of unknown ethnicity.

Source: Reprinted from *HIV-Related Tuberculosis*, p. 13, Centers for Disease Control and Prevention, 1992.

The Public Health Paradigm 227

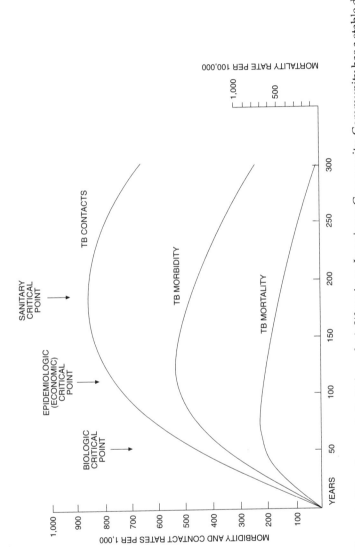

Figure 12-1 Theoretical Development of a Tuberculosis Wave in an Imaginary Community. Community has a stable degree of urbanization, and a population completely isolated from the rest of the world, in which tuberculosis is assumed to appear for the first time at moment zero. Shown are the death rate (inscribed on a different scale, at right), the rate of sickness, i.e., tuberculin-positive persons, which avoids complicated definitions, and the rate of contacts, all three by reference to living population. *Source:* Reprinted from E.R.N. Grigg, The Arcana of Tuberculosis, *The American Review of Tuberculosis and Pulmonary Diseases*, Vol. 18, No. 2, p. 160, 1958.

gered behind this is a curve representing the larger number of individuals who are infected with tuberculosis, but do not have clinical disease. This longer curve represents the fact that the vast majority of infected people (90 percent) never become clinically ill from the tuberculosis infection. Finally, this curve is followed still later by an ever greater number of contacts to the earlier cases. This curve represents the low communicability of tuberculosis compared to other contagious diseases such as influenza and measles. Thus, initial incident cases portend greater mortality and morbidity over time unless public health officials institute broad screening and treatment interventions. The use of public health data in constructing such curves is essential for tracking the natural history of such epidemics.

Different regions of the world are in different stages of these curves. In Europe, for example, the curve began centuries ago and has nearly run its course. The epidemic in North America, which was introduced much later, is still in its downward phase.[21]

Adjustments in the "closed" population, however, can alter an epidemic curve. For example, tuberculosis produces a more aggressive pattern in urban than in rural areas (Figure 12–2). Population density and the greater transmission of tuberculosis in crowded urban settings are responsible for this pattern. By accelerating the spread of tuberculosis, urbanization shifts the epidemic curve. Thus, all three constituent curves shift to the left, compressing greater numbers of infections into a more condensed time period. Compared to the traditional curves (see Figure 12–1), the three curves altered by urbanization reach their zenith well within the first 100 years, approximately half the amount of time required for the predicted effects in a rural setting.[22]

THE TUBERCULOSIS CURVE IN THE AGE OF AIDS

The construction of tuberculosis time trends in the mid-1980s revealed that something unexpected was occurring in tuberculosis case rates. For decades, these rates had been declining in the United States. Beginning in 1985, however, annual tuberculosis case rates in the United States began to increase. So consistent have been these increases that, overall, approximately 39,000 more cases of tuberculosis occurred in the United States between 1985 and 1991 than would have occurred if prior declines had persisted (Figure 12–3). As of 1993, the number of excess cases since 1985 has been estimated to be more than 63,000.[23] Such findings countered the long-standing decline in tuberculosis and challenged public health authorities to explain this unanticipated turn of events in an age of advanced medicine and public health. Why were data contradicting the predicted tuberculosis curve?

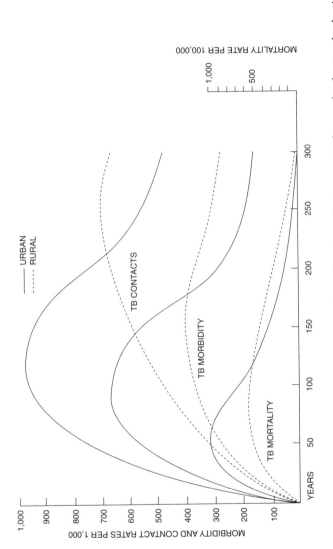

Figure 12–2 Tuberculosis Wave in Urban versus Rural Areas. Since the tuberculosis wave is the result of natural selection, its speed is proportional to the number of deaths per unit of time, and therefore, at any moment, the decrease will be commensurate with the preceding death rate. The pace of the wave is hastened by urbanization, because it increases the number of new contacts and raises the reactability potentials of exposed and afflicted persons (through stress and consequent alteration of corticosteroid secretions). The curves represent the theoretical tuberculosis wave in two extreme settings (one urban, the other rural), these two imaginary communities being assumed to remain isolated from the rest of the world and to maintain a constant level of urbanization throughout the three centuries shown. *Source:* Reprinted from E.R.N. Grigg, The Arcana of Tuberculosis, *The American Review of Tuberculosis and Pulmonary Diseases*, Vol. 18, No. 2, p. 162, 1958.

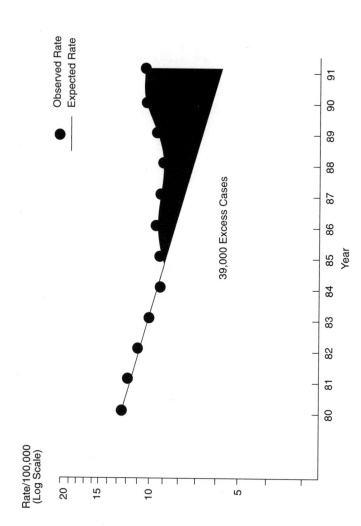

Figure 12-3 Expected and Observed Tuberculosis Cases in the United States, 1980–1991. *Source:* Reprinted from *HIV-Related Tuberculosis*, p. 11, Centers for Disease Control and Prevention, 1992.

Several events had transpired to alter the curve. The great declines of prior decades resulted in a large number of susceptible individuals. In New York and San Francisco in 1994, for example, molecular studies showed that between 30 and 40 percent of all *M. tuberculosis* infections were newly acquired.[24,25] More important, however, the advent of the HIV epidemic had altered individual immunology and produced a "hypersusceptible" population, facilitating reactivation of old infection and acquisition of new infection.[26] Once infected with *M. tuberculosis*, persons with HIV infection sustained a substantial risk of developing disease. These factors have important implications for tuberculosis control. To further hamper tuberculosis control, public health agencies had fallen victim to major budget cuts (Figure 12–4).[27] Federal categorical funding for tuberculosis control programs, the chief source of local and state control activities, including essential case finding, contact investigation, and screening, were replaced by flexible block grants. This permitted states to spend traditional tuberculosis funds on other public health needs. These major funding reductions for tuberculosis prevention services were believed to contribute to the resurgent tuberculosis epidemic. In addition, critical tuberculosis research activity waned, producing a gap in new drug development and diagnostic technology. (The last antimicrobial drug approved for the specific treatment of tuberculosis was rifampin in 1972.)

Old factors also persisted. In particular, the complex and lengthy period of treatment (i.e., the reliance upon at least four-drug therapy over a minimum of 6 months) fostered incomplete treatment and drug resistance.[28,29] Together, these factors challenged the public health system's capacity to respond to this emerging crisis.

Of these factors, the HIV epidemic has had perhaps the greatest influence on the re-emergence of tuberculosis. Early in the HIV epidemic, the link between HIV and tuberculosis became apparent to many clinicians and public health officials in Florida and New Jersey.[30,31] Ecological data soon supported this observation. Public health surveillance revealed that states and cities with the greatest number of acquired immunodeficiency syndrome (AIDS) cases were also the states and cities with the greatest number of tuberculosis cases. In point of fact, the top five states for incident tuberculosis were also the top five states for AIDS cases (Table 12–2). Moreover, in New York City, for example, the incidence of tuberculosis had paralleled the national decline prior to 1980, but increased some 140 percent between 1980 and 1991.[32,33] This increase resulted in nearly 15,000 more cases in New York City than would have occurred if prior declines persisted. Cross-matching public AIDS and tuberculosis case registries even more firmly established the AIDS-tuberculosis link, as it revealed

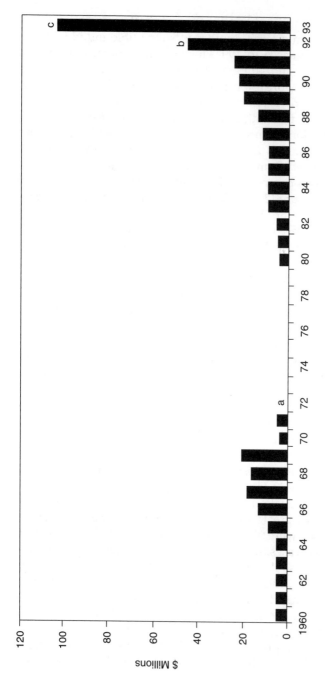

Figure 12–4 Tuberculosis Funding, U.S. Centers for Disease Control and Prevention, Fiscal Years 1960–1993. *Source:* U.S. Congress, Office of Technology Assessment, *The Continuing Challenge of Tuberculosis*, OTA-H-574, U.S. Government Printing Office, September 1993.

[a] Fiscal years 1972 through 1982 categorical grants ceased, funds to states were in block grants not specific for tuberculosis (TB).
[b] Fiscal year 1992 includes $26 million in human immunodeficiency virus (HIV) funds, used for HIV-related TB activities.
[c] Fiscal year 1993 includes $25 million in HIV funds used for HIV-related TB activities.

that approximately 4 percent of AIDS cases were in fact co-infected with tuberculosis, a rate substantially higher than that for the general population.

These findings led to targeted epidemiological studies in a search for possible high-risk groups. In New York, for example, 14 percent of tuberculosis patients with HIV infection, compared with 0 percent of tuberculosis patients without HIV infection, developed pulmonary tuberculosis over the 2-year study period. From such population-based data, it has been estimated that the risk of developing tuberculosis in persons with HIV infection is approximately 8 percent *per year*.[34] This stands in stark contrast to the risk to individuals without HIV infection, for whom the risk of developing tuberculosis is between 5 and 10 percent over their entire *lifetime*.

Additional public tuberculosis surveillance activities during the 1980s revealed a variety of new high-risk groups for tuberculosis, and policy recommendations were developed to provide greater screening of these groups:

- patients with co-existent HIV infection. The risk of developing tuberculosis in patients with HIV infection is 8 percent per year.[35]
- homeless populations. As many as 50 percent of homeless people in some areas may have *M. tuberculosis* infection, while some 7 percent may develop the disease.[36,37]
- prison populations. Tuberculosis infection among prison inmates is estimated to be four times that of the noncorrectional population.[38]
- foreign-born individuals from selected countries. Approximately 25 percent of all new tuberculosis cases occur among immigrants.[39]

Table 12–2 States Reporting the Most AIDS Cases and the Greatest Increase in Tuberculosis Cases

States	Rank by Number of AIDS Cases Reported through Dec.1991	Rank by Increase in Reported TB Cases 1985–1990
New York	1	1
California	2	2
Florida	3	4
Texas	4	5
New Jersey	5	3

Source: Reprinted from *HIV-Related Tuberculosis*, p. 21, Centers for Disease Control and Prevention, 1992.

Thus, the development of population-specific databases has been crucial to understanding several aspects of tuberculosis.

PUBLIC HEALTH DATA AND PROGRAM EVALUATION: A CASE STUDY

When used to evaluate public health program efficacy, ecological data can be useful to officials in developing public policy. Baltimore City's tuberculosis control program provides a case in point.

In 1978, the Baltimore City Health Department initiated a limited program of supervised therapy—Directly Observed Therapy (DOT)—for tuberculosis cases within Baltimore's municipal jurisdiction. DOT was implemented to counter the age-old problem of treatment nonadherence, which causes a significant risk of disease relapse and drug resistance. Initially, under this public program individual patients who were predetermined to be most likely not to complete therapy (e.g., the homeless and drug abusing patients) were offered full and free treatment through the Baltimore City Health Department. Participating patients received nurse-supervised DOT in one of the city's tuberculosis clinics for the duration of treatment.

In 1982, this program was expanded to make all tuberculosis patients in Baltimore eligible for treatment through the city's DOT program. Over the ensuing decade, more and more patients received treatment under DOT, and policy makers began to explore the impact that this program was having on Baltimore's tuberculosis problem. With the use of public disease control data, several indirect measures of this program's effectiveness were developed.

In contrast to the increased incidence in most other major metropolitan areas, the incidence of tuberculosis declined significantly during the 1980s in Baltimore. Compared to the national large city increase of 10 percent, the incidence declined nearly 65 percent in Baltimore. Baltimore's ranking for tuberculosis incidence reflects this decline. During the 1960s and 1970s, Baltimore routinely ranked among the top two or three cities for incident tuberculosis (number 1 in 1975). Based on tuberculosis case rates, Baltimore's ranking fell to 28th by 1992[40] (Figure 12–5). Other publicly collected data partially explained these trends. Patients who were receiving DOT were more likely than patients who were administering their own therapy to complete treatment. According to public health surveillance data, on an average, more than 90 percent of Baltimore's DOT patients completed their treatment, compared to a national average of around 76 percent.[41] Furthermore, a more sensitive measure, the rate of sputum con-

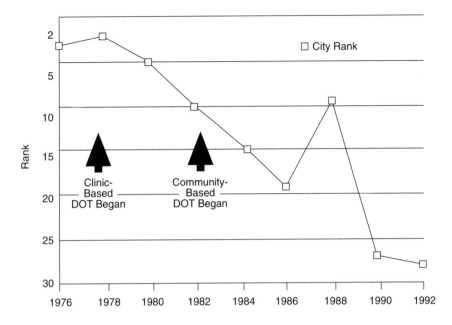

Figure 12-5 Baltimore's National Ranking for Tuberculosis Incidence among U.S. Cities over 250,000 Population, 1976–1992. *Source:* Reprinted from C.P. Chaulk et al., Treating Multidrug-Resistant Tuberculosis: Compliance and Side Effects, *Journal of the American Medical Association*, Vol. 271, No. 2, p. 104, the American Medical Association, 1994.

version (i.e., the point at which sputum specimens no longer grow *M. tuberculosis*, the patient is considered noncontagious, and cure, if therapy is completed, is expected), for DOT patients was generally twice that of private sector patients.

CONCLUSION

Health care policy is frequently viewed as a formative process based on data extracted from such medical encounters as physician visits, hospital discharges, or per capita expenditures. Health care policy decisions with respect to public health often rest on ecological or population-based data sets. By necessity, these analyses are not as robust as the traditional clinical trial or case control study conducted in clinical medicine.

However, regardless of which paradigm is employed, health care policy decisions have similar goals: to reduce morbidity and mortality, and to deploy resources in the most economical fashion toward these goals. Generally, private sector policy focuses on the individual and uses discrete events in time (i.e., encounters with the health care system) as the unit of measure from that medical event and resources consumed. In public sector health issues such as the spread of communicable disease, these discrete events are summed for a population over a selected period of time and are commonly reported as disease incidence.

The ability of a society to develop rational strategies for solving public health problems rests on the development of unique population-based data. Determining disease trends, defining populations at risk and the natural history of a disease, and conducting programmatic evaluations require a valid and accessible population data set. Without such data, neither the medical care system nor the public health system can economically and effectively direct their resources to improve the public's health status.

To assess the performance of a health care system, policy makers must go beyond data such as the number of physician visits or the percentage of the population without health insurance. An important measure of a health care system's performance is the health status of the population. The challenge of health care policy makers is to institute systemic changes that improve the public's health status. Alterations in such traditional core public health services as tuberculosis control may be jeopardized, however, unless new systems of care maintain such critical public services as disease surveillance, case finding, contact investigation, screening of high-risk groups, and chemoprevention. Assessing the impact of systemic change on these core public health functions will require the continued collection, assessment, and interpretation of population-based data. A failure to appreciate the role of such data could lead to false or irresponsible policy decisions, squander precious disease control resources, and jeopardize the public's health status.

NOTES

1. Institute on Medicine, Committee on the Study of the Future of Public Health National Academy of Sciences, *The Future of Public Health* (Washington, D.C.: National Academy Press, 1988).
2. W. Cates, Jr., and A. Meheus, Strategies for Development of Sexually Transmitted Diseases Control Program, in *Sexually Transmitted Diseases*, ed. K.H. Holmes (New York: McGraw Hill, 1984).
3. R. Dubos and J. Dubos, *The White Plague: Tuberculosis, Man, and Society* (New Brunswick, N.J.: Rutgers University Press, 1992).

4. F. Ryan, *The Forgotten Plague: How the Battle Against Tuberculosis Was Won—And Lost.* (Boston, Mass.: Little, Brown & Company, 1993).
5. Centers for Disease Control and Prevention, *Tuberculosis Statistics in the United States 1992* (Atlanta, Ga.: U.S. Department of Health and Human Services, Public Health Service, 1992).
6. A.B. Bloch et al., The Epidemiology of Tuberculosis in the United States: Implications for Diagnosis and Treatment, *Clinics in Chest Medicine* 10 (1989):297–313.
7. R. Koch, Die Aetiologie der Tuberculose, *Berliner Klinische Wochenschrift* XIX (1882):221.
8. B.E. Strong and G.P. Kubica, *Isolation and Identification of Mycobacterium Tuberculosis* (Atlanta, Ga.: U.S. Department of Health and Human Services, Public Health Service, CDC, 1981), DHHS Publication Number (CDC) 81-8390.
9. F.C. Tenover et al., The Resurgence of Tuberculosis: Is Your Laboratory Ready? *Journal of Clinical Microbiology* 31 (1993):767–770.
10. G. Canetti et al., Micobacteria: Laboratory Methods for Testing Drug Sensitivity and Resistance, *Bulletin of the World Health Organization* 29 (1963):565–578.
11. J.A. Califano, Three-Headed Dog from Hell: The Staggering Public Health Threat Posed by AIDS, Substance Abuse, and Tuberculosis, *Washington Post* December 22, 1992:A22.
12. S.H. Ferebee, Controlled Chemoprophylaxis Trials in Tuberculosis: A General Review, *Advances in Tuberculosis Research* 17 (1970):1–6.
13. P.M. Small et al., The Epidemiology of Tuberculosis in San Francisco: A Population-Based Study Using Conventional and Molecular Methods, *New England Journal of Medicine* 330 (1994):1703–1709.
14. D. Alland et al., Transmission of Tuberculosis in New York City: An Analysis by DNA Fingerprinting and Conventional Epidemiologic Methods, *New England Journal of Medicine* 330 (1994):1710–1716.
15. Small et al., The Epidemiology of Tuberculosis in San Francisco.
16. S.C. Etkind, The Role of the Public Health Department in Tuberculosis, *The Medical Clinics of North America* 77 (1993):1303–1314.
17. J.F. Broekmans, Evaluation of Applied Strategies in Low-Prevalence Countries, in *Tuberculosis: A Comprehensive International Approach*, ed. L.B. Reichman and E.S. Hershfield (New York: Marcel Dekker, Inc., 1993), 641–667.
18. J.L. Hadler, Control of Tuberculosis, in *Tuberculosis: Current Concepts and Treatment*, ed. L.N. Friedman (Ann Arbor, Mich.: CRC Press, 1994), 307–333.
19. J.A. Jereb, G.D. Kelly, and S.W. Dooley, Tuberculosis Morbidity in the United States: Final Data, 1990, *Morbidity and Mortality Weekly Report* 40 (1991):23–27.
20. E.R.N. Grigg, The Arcana of Tuberculosis: With a Brief Epidemiologic History of the Disease in the U.S.A., *The American Review of Tuberculosis and Pulmonary Diseases* 18 (1958):151–172.
21. J.H. Bates and W.W. Stead, The History of Tuberculosis As a Global Epidemic, *Medical Clinics of North America* 77 (1993):1205–1217.
22. Grigg, The Arcana of Tuberculosis.
23. Expanded Tuberculosis Surveillance and Tuberculosis Morbidity–United States, 1994. *Morbidity and Mortality Weekly Report* 43 (1994):361–366.
24. Small et al., The Epidemiology of Tuberculosis in San Francisco.
25. Alland et al., Transmission of Tuberculosis in New York City.

26. C.L. Daley, P.M. Small, and G.F. Schechter, An Outbreak of Tuberculosis with Accelerated Progression among Persons Infected with the Human Immunodeficiency Virus: An Analysis Using Restriction-Fragment-Length Polymorphisms, *New England Journal of Medicine* 326 (1992):231–235.
27. U.S. Congress, Office of Technology Assessment, *The Continuing Challenge of Tuberculosis*, OTA-H-574 (Washington, D.C.: U.S. Government Printing Office, September, 1993).
28. J.A. Sbarbaro, Public Health Aspects of Tuberculosis: Supervision of Therapy, *Clinics in Chest Medicine* 1 (1980):253–263.
29. J.A. Sbarbaro, The Patient-Physician Relationship: Compliance Revisited, *Annals of Allergy* 64 (1990):325–331.
30. A.E. Pitchenik, C. Cole, and B.W. Russell, Tuberculosis, Atypical Mycobacteriosis, and the Acquired Immunodeficiency Syndrome among Haitian and NonHaitian Patients in South Florida, *Annals of Internal Medicine* 101 (1984):641–645.
31. G. Sunderam, R.J. McDonald, and T. Maniatis, Tuberculosis As a Manifestation of the Acquired Immunodeficiency Syndrome (AIDS), *Journal of the American Medical Association* 256 (1986):362–366.
32. Centers for Disease Control and Prevention, *Tuberculosis Statistics in the United States 1980* (Atlanta, Ga.: U.S. Department of Health and Human Services, Public Health Service, 1980).
33. Centers for Disease Control and Prevention, *Tuberculosis Statistics in the United States 1991* (Atlanta, Ga.: U.S. Department of Health and Human Services, Public Health Service, 1991).
34. P.A. Selwyn et al., A Prospective Study of the Risk of Tuberculosis among Intravenous Drug Users with Human Immunodeficiency Virus Infection, *New England Journal of Medicine* 320 (1989):545–550.
35. Ibid.
36. Tuberculosis Control among Homeless Populations, *Morbidity and Mortality Weekly Report* 36 (1987):257–260.
37. C.W. Schieffelbein and D.E. Snyder, Jr., Tuberculosis Control among Homeless Populations, *Archives of Internal Medicine* 148 (1988):1843–1846.
38. Centers for Disease Control and Prevention (Atlanta, Ga.: U.S. Department of Health and Human Services, Public Health Service, unpublished data).
39. Bloch et al., The Epidemiology of Tuberculosis in the United States.
40. C.P. Chaulk et al., Treating Multidrug-Resistant Tuberculosis: Compliance and Side Effects, *Journal of the American Medical Association* 271 (1994):103–104.
41. Centers for Disease Control and Prevention, *Tuberculosis Program Management in the United States 1986–1991* (Atlanta, Ga.: U.S. Department of Health and Human Services, Public Health Service, March 3, 1993).

» Chapter 13 «

Measurable Accountability in an Era of Health Care Reform

Kathleen R. Ciccone and Onita D. Munshi

THE 1990s are proving to be a watershed period for health care reform. Although a variety of ideas and plans have surfaced, the consensus is that no single agenda is likely to succeed. The final version of health care reform will be a blend of many ideas, with incremental insurance reform as a likely first step. Despite the diversity in approaches, however, a number of elements are common to most of the reform plans put forward.

The first common element—and a primary force behind the drive for health care reform—is broad access to care. Although health expenditures consume more than 13 percent of the gross national product,[1] many individuals in the United States are without legitimate access to basic services. Spiraling costs for employers (as purchasers of health care coverage for employees) and their impact in a competitive global economy have served as a "wake-up" call for all purchasers, who have begun to explore and insist on improved value for their health care dollars.

Second, there is concern about the lack of information available for making individual health care decisions, regarding both what service(s) may be needed and who should provide those services. There are considerable variations in approaches to health care delivery, but no one approach has proved to be most effective. Given the dearth of reliable information, how do individuals make informed, intelligent decisions about their own or their family's health care?

The authors are indebted to Edward L. Hannan, PhD, State University of New York at Albany, who provided much of the data and statistical references for the chapter, and also to him and several physicians and hospital representatives who took the time to review the text and provide comments. However, all opinions expressed herein are the sole responsibility of the authors.

A third element common to all health care reform agendas is the perception that the dissemination to the public of information about provider performance is a viable strategy for increasing the public accountability of health care providers. Three assumptions underlie this strategy: (1) that the broadcast and examination of provider-specific information (score cards or report cards) through the local media or other public avenues will stimulate health care providers to address and improve both the quality and efficiency of care; (2) that adequate measures of quality are available and can be collected, making it possible to assess value; and (3) that patients and purchasers will use this information to make decisions about providers and treatment options.

Regardless of whether available data are valid or any of these assumptions are accurate, the grades are coming in. Until 1992, for example, the Health Care Financing Administration and Veterans Affairs Hospitals publicly released annual statistics on mortality, including rates for coronary artery bypass surgery. Some individual states, accrediting bodies, and peer review organizations have initiated their own systems for measuring and reporting on the performance of health care providers. Such performance systems include the Indicator Measurement System of the Joint Commission on Accreditation of Healthcare Organizations, the Health Plan Employer Data and Information Set (HEDIS) of the National Committee for Quality Assurance, and the coronary artery bypass graft (CABG) surgery cost and outcomes reporting system of the Pennsylvania Health Care Cost Containment Council.

Questions and concerns abound. Are public report cards appropriate? Can the salient features of health care delivery be measured and communicated accurately to the public? Will the public really use this information? It is too early to answer most of these questions; the collection of evidence is just beginning. Clearly, it is appropriate for communities, patients, and purchasers to have easy access to information that will help them make decisions about their own health care. It is their choice whether to use this information. Given the enormity of the fallout from public score cards, however, providers are justifiably concerned that information about them and their practices be accurate, valid, and placed in the proper context. Current measures of quality are undeniably both immature and incomplete, and their usefulness for consumers is still unknown.

Although providers may believe that the lay public and purchasers are unable to interpret measures of quality accurately, Justice Harold J. Hughes took a different approach in his decision in the matter of *Newsday, Inc. vs. the New York State Department of Health*.[2] David Zinman, a journalist from *Newsday*, had sought to obtain physician-specific patient mortality data from the New York State Cardiac Surgery Reporting System (CSRS).

The New York State Department of Health (NYSDOH) had already published hospital-specific data. The NYSDOH refused Zinman's request, citing protection of the physicians' privacy and concern over the public's ability to interpret the data accurately. In his October 15, 1991 decision, Justice Hughes chided the NYSDOH for believing that "the State must protect its citizens from their intellectual shortcomings by keeping from them information beyond their ability to comprehend. . . ." He continued, "The duty of administrators to release to the public the records of its government cannot be dependent upon the administrators' assessment of the population's intelligence."[3] Regarding physicians' expectation of privacy about the results of their work, Justice Hughes wrote, "Furthermore, even if there was a legitimate privacy expectation, the interest of the public outweighs it."[4] In this precedent-setting case, the court recognized the needs of various publics (e.g., private citizens, insurance companies, government) and the legitimate expectation for access to information that would enable patients to make more knowledgeable health care decisions.

Public access to data is a current fact of life, and the challenge is to craft a forum for accountability that recognizes the power of such information and respects the needs of the providers of health care, as well as the recipients of the data.

NEW YORK STATE CARDIAC SURGERY REPORTING SYSTEM

Considered "precedent-setting" (with all the attendant risks and benefits) in the provider report card field, the NYSDOH initiated a cardiac surgery mortality data collection and reporting system that makes selected performance statistics available to the provider community and to the general public. In 1989, the NYSDOH began collecting clinical data prospectively on all patients undergoing open heart surgery, specifically CABG surgery. All hospitals in the state that are approved to perform cardiac surgery provide these data to the CSRS. The major purposes of the CSRS are (1) to provide hospitals with information that will assist them in assessing and improving their quality of care and in determining when cardiac surgery is appropriate; (2) to assist the NYSDOH in its quality improvement activities; and (3) to provide consumers with information that will aid them in selecting providers of cardiac surgery.[5] The New York State Cardiac Advisory Committee, which consists primarily of independent practicing cardiologists and surgeons from New York and other states, oversees the program.

The NYSDOH has used the registry information to develop a statistical model that determines which preoperative risk factors in the data system are significantly related to in-hospital death, and predicts the probability of adverse outcomes, given the presence of various risk factors. When adjusted for the severity of patients' preoperative conditions, the resulting information should allow an assessment of the performance of surgeons and hospitals over time by permitting a comparison of a predicted mortality rate for each with an actual or crude mortality rate.

For each patient who undergoes CABG surgery, the hospital's cardiac surgery department completes a form that contains patient demographic information; hospital and surgeon identifiers; preoperative risk factors; complications during surgery or postsurgical care; dates of admission, discharge, and surgery; and discharge disposition (alive or dead). This information is transferred to a personal computer by means of data entry software that checks for various data entry omissions, errors, and inconsistencies. The data are sent on diskette each quarter to the NYSDOH, which acts as the data-coordinating and analysis center.[6]

The NYSDOH thought it necessary to determine which of the preoperative risk factors contained on the form were significant predictors of adverse outcomes and to weight these risk factors in order to predict any given patient's risk of an adverse outcome. For example, an 80-year-old patient with diabetes and a history of myocardial infarction would have a higher risk of dying in the hospital than a 40-year-old patient with no significant risk factors. Hospitals are provided with a list of preoperative risk factors for CABG surgery that are significantly related to inpatient mortality. Hospitals then mark the risk factors present in a given patient to assess the probability of the patient's in-hospital death.

The data collection and analysis system adjusts the outcomes for each hospital to account for differences in the average patient severity of illness in order to ensure that hospitals treating the most severely ill patients are not unfairly penalized for higher actual mortality rates. This risk-adjusted mortality rate is an estimate of what a hospital's mortality rate would have been if its patients had exactly the same risk factors as did the patients in all other hospitals. Each hospital is given the software that enables the hospital to calculate the risk-adjusted mortality rates for each of its surgeons for any specified period during the year.[7]

To examine the changes that have occurred in outcomes for CABG surgery in New York State in the period 1989 through 1992, the NYSDOH studied:

1. significant risk factors related to in-hospital mortality for the 4-year period
2. changes in overall outcomes (risk-adjusted mortality rates), patient

preoperative severity of illness (expected mortality rates), and volumes in the same 4-year period
3. hospital-specific changes in risk-adjusted mortality rates and volumes in the 4-year period
4. the relationship between hospital risk-adjusted rates and the average severity of illness of patients undergoing surgery
5. the ability of the statistics to predict mortality as a function of the severity of illness of a case[8]

The results indicated that the volume of isolated CABG operations in New York State rose 30.6 percent between 1989 and 1992. This increase was accompanied by a 21 percent decrease in actual in-hospital mortality rates, from 3.52 to 2.78 percent in the 4-year period. The actual mortality rate decreased during the period, although it has been suggested that the average severity of illness of patients undergoing surgery may have increased, as evidenced by the increase in expected mortality rate from 2.62 percent in 1989 to 3.54 percent in 1992. The results also demonstrated that, after adjustments for patient severity of illness, the mortality rate for isolated CABG surgery decreased from 4.17 percent in 1989 to 2.45 percent in 1992, a decline of 41 percent. Although no directly comparable data are available from other states, New York's mortality rate is the lowest so far reported in the United States. A total of 27 of the 30 hospitals performing cardiac surgery in the state experienced a reduction in their risk-adjusted rate between 1989 and 1992.

Based on the analysis of the registry data, the Cardiac Advisory Committee and the NYSDOH have conducted site visits to hospitals whose cardiac surgery programs have needed special attention. They have recommended that some facilities obtain outside consultants to improve their cardiac surgery programs and have placed some programs on probation until recommended changes could be instituted. Some programs replaced surgeons, refined patient selection criteria, evaluated the condition of patients more closely for preoperative risk, and directed patients to more appropriate surgeons.

The NYSDOH has hypothesized, but not proven, that a large part of the decrease in both actual and risk-adjusted mortality rates resulted from improvements in the quality of care emanating from New York State's new system, as many hospitals have modified patient practices or hired new chiefs of cardiac surgery programs. There are other possible explanations for these changes, however: (1) similar, but less dramatic, mortality rate reductions occurring nationally, which may reflect improvements and advancements in surgical techniques and postoperative care; (2)

changes in both patient risk (known to be increasing nationally), and in thoroughness of risk factor documentation and data collection; (3) physicians' referring high-risk patients in an effort to improve or protect their own performance assessments.

Consumer information for CABG surgery has been available since 1989, when the NYSDOH began publicly releasing consumer information booklets on the volume of cases, crude mortality rates, and risk-adjusted mortality rates by hospital. These booklets identify which hospitals have risk-adjusted mortality rates that are significantly higher or lower than the statewide rates for CABG surgery. The NYSDOH released surgeon-specific information after losing the lawsuit filed in 1991 by *Newsday*. Currently, the NYSDOH voluntarily releases surgeon-specific, risk-adjusted mortality rates for CABG surgery for surgeons who have performed at least 200 isolated CABG operations in a 3-year time period. This minimum volume of 200 operations for the 3-year period was chosen because smaller numbers of operations do not yield statistically significant results. The NYSDOH consumer information booklets also identify which surgeons have rates that are higher or lower than statewide rates and encourage prospective patients and their families to discuss the data with their physicians and to ask questions about their own risk profiles generated by the computer software.

New York has one of the few reporting programs that publicize hospital- and surgeon-specific data. Obviously, the press has generated a great deal of interest in these data. Although the NYSDOH had never intended to rank-order hospitals according to mortality statistics, the media has, as evidenced by one of the early headlines, which read, "State Ranks Cardiac Hospitals."[9]

AUDIENCE FOR THE CSRS DATA

As indicated earlier, the data collection and analysis of cardiac surgery outcomes, specifically mortality rates, began for the purpose of quality improvement. Thus, two of the primary audiences for the data are hospitals and cardiac surgeons. The NYSDOH's quality improvement activities emanating from the CSRS data are directed at these two interdependent providers. The goal of providing consumers with information has been a driving force in the increasing demand for provider report cards. Consumers and purchasers of health care (e.g., employers, insurers), as well as state governments, have insisted on the public release of data about provider performance and efficiency in order to make providers accountable for the quality of the care that they provide.

There are similarities in some of the reactions of the hospitals and surgeons to the publication of the CSRS data, as well as concerns and responses that are individual to each group. Patient response has differed from what might have been expected in many cases.

The Hospital Experience

When the figures for hospitals were publicly released by the NYSDOH for the first time in 1992, reactions were strong. Although the early releases cautioned consumers not to base any health care decisions solely on the data, hospitals with unfavorable rates were understandably concerned that the information could change and adversely affect the appeal of their programs. The reliability and validity of the data were questioned. Hospitals with good ratings were concerned that they would have an influx of very high-risk patients that, they feared, could have a negative impact on their rates in future reports. They also questioned the sensitivity of the data-reporting system.

Results of a follow-up study by the NYSDOH suggested that the expected widespread referral of the more difficult CABG cases to hospitals with superior ratings did not occur, at least not at the time of the study.[10] Anecdotal information and opinions gathered in discussions with some hospital representatives suggest otherwise, however, indicating that further investigation is warranted.

One pattern that was consistent throughout the first 4 years of the CSRS was the relationship between case volume and mortality rates. A high volume of CABG surgeries was generally associated with a lower mortality rate. Cardiac surgery departments in some hospitals responded to that information in a number of ways. In some instances, hospitals excluded surgeons who had had a lower surgery case load in the past. In other instances, hospitals required surgeons to perform a minimum number of operations annually. Some cardiac departments changed their patient acceptance patterns, referring patients at high risk to other hospitals or to more experienced surgeons within the hospital. Some hospitals began to require newer, less experienced physicians to assist experienced surgeons during more procedures until they achieved a higher level of competence. In a few hospitals, the chief of staff or head of cardiac surgery was replaced, and/or entire programs were reorganized.

One hospital's evaluation process revealed an increased mortality rate associated with emergency surgeries and very high-risk patients. The staff response was to change their practice to improve the stabilization of patients prior to surgery and, in the process, channel these very high-risk patients to the most experienced and most skilled surgeons.

The Physician Experience

The release of physician-specific performance data in response to the *Newsday* decision in 1991 was threatening to many surgeons, who claimed that having rates that were higher than the statewide average—whatever the reason—could potentially destroy their reputations and careers. They expressed concerns that the data unfairly reflected circumstances beyond their control and that the risk adjustment methodology would not adequately compensate for the presence of multiple high-risk patients in their practice. They also felt that the adjustment was inadequate for the random variation of the "luck of the draw."

One of the first problems that became apparent with the physician-specific data was that the numbers for most physicians within a single year were insufficient to yield meaningful data. Although the NYSDOH was confident that the risk adjustment process fairly and accurately compensated for differences in patient severity of illness over a large volume of cases, it could not accurately predict future performance based on the low volume of cases available for surgeons in the first few years of the program. An examination of the data's ability to predict future physician performance based on 1 year's rates showed that ability to be only slightly better than pure chance. The current data are somewhat more meaningful, because, as mentioned earlier, ratings are released to the public only for surgeons who performed at least 200 operations during a given 3-year period.

Some physicians have predicted that the physician report cards may delay the development of new technologies. The rationale for this concern is that errors may be made when surgeons are learning new procedures. Errors can translate into higher mortality rates on the report cards, discouraging surgeons from practicing and using new techniques. It is difficult to assess whether or to what degree this may be occurring.

Changes in hospital policies regarding who can operate, the regulations regarding minimum numbers of surgeries, and increased training requirements affect nearly all surgeons, but particularly those whose performance ratings vary significantly from the statewide average. Like the hospitals, physicians are gradually accepting the public nature of the data and making a proactive attempt to use the data to improve care. Some surgeons have even begun posting the ratings visibly in their offices. The data over the first 4 years of this program have demonstrated dramatic improvement in mortality rates, and there is an underlying assumption that these rate improvements are a reflection of overall improvements in the quality of care of cardiac surgery patients.

The Patient Experience

When the hospital and physician ratings first went public, before there were enough years' data to provide a truly meaningful picture of a provider's performance over time, access was a concern for providers and consumers alike. Lingering questions about the validity of the methodology for risk factor adjustment have continued to fuel this concern. The NYSDOH 1992 study and results of ongoing analyses of the data have implied that providers are not turning away high-risk patients in significant numbers, however, in order to protect their report cards.[11] When a high-risk patient is referred, it is to providers who are doing a high volume of surgeries and maintaining a satisfactory performance rating. Ultimately, such a referral should improve the referred patient's chances for a satisfactory outcome, but this theory needs further study. Anecdotal reports have suggested different conclusions and require additional exploration.

Although consumer advocacy groups were among the most vocal in demanding that outcome data be made public so that consumers could make informed decisions about health care and health care providers, consumers do not appear to be taking full advantage of the information in the manner anticipated or intended. It was expected that the data would be helpful in framing questions and, ultimately, in making choices regarding where and from whom to seek care. This use of information was not reflected in practice; however, patients did not flock to physicians and hospitals with the best rankings.

There could be a number of reasons for this. Many patients may find it difficult to question a physician whom they trust and depend on for their health care needs. Even if a patient does not have a long-standing relationship with a cardiac surgeon, he or she may often rely on referrals from a cardiologist. Once again, conversations suggest that cardiologists have not generally changed their referral patterns, perhaps because of their concerns about the limitations of the data and the rankings. The lack of statewide systematic approaches for educating patients about the cardiac surgery reports may also contribute to patients' limited use of the information.

Even if a patient takes the initiative to research the mortality rates of the surgeon recommended to him or her, there is no assurance that a surgeon with a favorable rate is a participating provider in the patient's health care plan. Although many plans make allowances for patients to go outside the plan for a particular provider, they usually require the patients to pay much more than they would pay for a participating provider. The difference between paying no deductible or a small percentage of the cost of

CABG surgery and paying 40 to 50 percent to go outside the plan can mean tens of thousands of dollars—not a figure that the average patient can ignore.

If an individual was not considering cardiac surgery at the time of the cardiac surgery report release, the individual may have forgotten about the availability of the data by the time that cardiac surgery is recommended. If the data could be prepared and available in a manner that is meaningful and timely for patients, and appropriate educational materials developed, then physicians, hospitals, and patients would be likely to feel more comfortable in actively using the data in their decision-making process.

Finally, awareness of the availability of the cardiac surgery mortality statistics is limited. Although there was significant publicity when the provider report cards first hit the media in 1991, much of the controversy has diminished. They are no longer front page news.

LESSONS LEARNED

Although New York State's data collection and quality improvement program has not been flawless, it has been a valuable example and has raised many questions, as yet unanswered, that could be instructive for national, state, and local initiatives that seek to use the CSRS as a template. From the perspective of New York State hospitals, the initial flaw in the system was probably the failure of the NYSDOH to recruit hospital administration representatives to participate in the dialogue on cardiac surgery. Although many of the physicians who were members of the Cardiac Advisory Committee practiced and ran programs in New York State hospitals, the clinical discussion that occurred did not always take into account the organizational perspective. The program was designed and run as a physician-driven process. Even though the first release of data focused on aggregate hospital mortality ratings, hospitals were largely uninformed and unaware of the goals of the program.

One of the primary goals of the cardiac surgery program was to assist the NYSDOH in quality improvement activities. For true quality improvement, it is necessary to view patient outcomes, surgical competence, and hospital processes collectively rather than in isolation. If hospital management had been represented at the table and had been actively involved in the CSRS, even greater improvements may have been achieved. Critical examination of care and redesign for quality initiatives require real changes that cannot occur without institutional buy-in to the project.

Shared ownership is strong motivation for active participation. Many hospitals viewed the initial data releases as a "regulatory club," however, and they responded in kind. They considered the data and motivations for the release to be suspect, and they raised questions about the program's methodology. Considerable reactionary energy was directed at pointing out flaws in a system that hospitals had no voice in developing. The state's data release may have directed hospitals' attention to their cardiac surgery programs as a "front burner" issue, but at the cost of a statewide partnership for care. Given the limited resources and the span of items that a single institution can prioritize as front burner issues, other, more pressing issues may have been relegated to the "back burner."

Variability in data reporting and ranking has clearly demonstrated the importance of frequent auditing and testing of the accuracy and validity of both the raw data and the resulting rates. The providers and the collector of the data share the onus of responsibility for the accuracy of the data. It is a hospital's responsibility to provide valid and accurate data. The collector, the NYSDOH, has an obligation to provide clear and consistent definitions of the data requested and to subject both the data and the analysis to frequent audits. The importance of testing and retesting the methodology increases when changes are made to any part of the system.

Organizers of ongoing data collection systems should hold frequent "users' group" meetings to allow participants to discuss and provide input regarding data collection and definition issues. Not only will such discussions lead to provider "ownership" of the data, but also frequent discussions may lead to increased reliability of data.

The media is a powerful source of information for the general public. Information must be packaged and presented to the media in a way that will benefit the public without causing unnecessary harm to hospitals, to physicians, and possibly even to patients. Initially, sensationalized headlines and sexy sound bites seem to capture the public's attention. As noted earlier, based on the first data release by the NYSDOH, the media rank-ordered hospitals according to their risk-adjusted mortality rates. Although there was little statistical difference in the findings, the media and public perception was that the care provided by a hospital ranked second was substantially superior to the care provided by a hospital ranked seventh. Ideally, if provider report cards are to serve the purpose of driving quality, the media needs to help educate the public about the meaning of the data and the value of using the data, not as a scorecard for choosing a provider, but as a resource for asking informed questions and making educated decisions.

The risk adjustment methodology is intended to be sensitive to differences in patient severity of illness, but there continues to be some question whether providers have restricted access to care for patients with more complicated conditions for fear of looking bad on the report cards. Conversely, some providers have made dramatic improvements in their program by initiating a more rigorous patient selection methodology that is appropriate for their level of expertise. Hospitals and physicians have begun to realize that they must perform an objective assessment of their own capabilities and limitations and that referring selected patients to more highly skilled providers is appropriate.

Hospital and provider rankings have created a competitive environment. Hospitals and physicians that recognize the need for technical and process quality improvements have sometimes sensed reluctance from others when they have reached out for guidance on best practice.

Although very complex, cardiac surgery is easier to examine than many other clinical treatments. Thus, it was probably a good place to start building a database. As a capitated system of networks develops, the appropriateness and effectiveness of other procedures and services, particularly those of high cost or high volume, will come under increased scrutiny. The NYSDOH has begun to explore some of these other areas, beginning with treatments for breast cancer. Having learned from the past, hospitals, physicians, and government representatives are making every effort to travel that path together.

UNRESOLVED ISSUES

As with any new initiative, a number of flaws have been identified in the cardiac registry program. The NYSDOH, with the guidance of the Cardiac Advisory Committee, has continued to refine and improve the CSRS to produce increasingly more meaningful data. However, some questions and concerns remain unresolved.

The system measures only in-hospital mortality. It does not gather data on the patient's survival following discharge. Furthermore, while mortality may be the easiest indicator to track, it may not be the most meaningful index of the quality of health care. Other outcomes, such as length of stay, readmissions, functional status, or recurrence of symptoms, contribute to the total quality picture. Death rates alone do not measure the quality of a provider's care.

Issues also remain regarding the variability of the data. In a recent audit of the CSRS data from a sample of ten hospitals, the data reported by one of the hospitals were found to be flawed. The data were corrected thereafter at that hospital. As the CSRS continues, the collection and analysis of the data, as well as the kinds of data to be included, should be reassessed on an ongoing basis. In addition, for samples to provide meaningful data from which future provider performance may be predicted, the sample size should be larger and should cover a longer span of time. Low numbers of patients or procedures for a given practitioner or the entry into practice of new physicians late in the data collection year can skew the statistics for some hospitals and/or individual physicians.

Volume plays a role in the performance rating of both hospitals and individual practitioners. According to the statistics, the higher volume providers consistently performed better. "Directly or indirectly, medical scorecards may create a negative feedback loop, so that hospitals or physicians with inadequate performance have decreasing clinical activity, with decreasing caseloads, which further decreases the technical competence of the physician or team involved."[12] Moreover, if providers with poorer patient outcomes were to be reimbursed at a lower rate (as some government and other third-party payers are proposing), this negative feedback loop would be compounded by the lack of investment in physicians who may benefit the most from further training and skill development.

Concerns regarding access persist, specifically the fear that some surgeons and hospitals would limit access to care for the most severely ill patients. This has not been reflected in the data, although anecdotal reports suggest a need for further study. According to the NYSDOH, increases in hospital volume have been independent of the published mortality rates. The methodology developed to adjust for patient severity of illness should not be a concern, in theory, except as physicians refer elsewhere because they have identified limitations in their own technical competence; such an event is not a negative impact. A review of the data reveals that, over a 4-year period, hospitals with more severely ill patients experienced better risk-adjusted outcomes. Here again, continuous testing of the methodology and validation of the accuracy of the impressions created by the data are necessary.

CONCLUSION

Generally, the New York State experience demonstrates the potential for providers and government representatives to work together to implement quality improvement programs that include public release of hospital- and physician-specific data. Some believe that, while better surgical technique and other factors unrelated to this program probably played a role in the improvements measured in outcomes of CABG surgery from 1989 to 1992, the collection and feedback of the data on risk-adjusted outcomes, the quality improvement activities they spawned, and their public release have been contributors to better health care.

The demand for measures of accountability of health care providers and the movement toward making the results of those measurement available to the public are growing. The development initiatives similar to the New

York State CSRS, but located in other geographical areas and focused on other health care issues, is inevitable. The challenge facing the industry is to draw on the expertise of multiple resources, with their sometimes conflicting agendas, to create systems that provide accurate and meaningful information in a format that meets the needs of a diverse audience. Experience with the cardiac surgery program in New York State will be invaluable as new ideas on the use of outcome data systems for assessing quality and efficiency in health care are explored.

NOTES

1. U.S. Health Care Financing Administration, Office of the Actuary, Data from the Office of National Health Statistics (Baltimore, Md.: 1991–1992).
2. *Newsday, Inc. vs. New York State Department of Health*, WL28S624 (Supreme Court of the State of New York, County of Albany), October 15, 1991.
3. Ibid.
4. Ibid.
5. E.L. Hannan et al., Improving the Outcomes of Coronary Artery Bypass Surgery in New York State, *Journal of the American Medical Association* 271, no. 19 (1994):761–766.
6. Ibid.
7. Ibid.
8. Ibid.
9. P. Wehrwein, State Ranks Cardiac Hospitals, *Albany Times Union*, December 5, 1990:A1, A3.
10. Hannan et al., Improving the Outcomes.
11. Ibid.
12. E.J. Topol and R.M. Califf, Scorecard Cardiovascular Medicine—Its Impact and Future Directions, *Annals of Internal Medicine* 120, no. 1 (1994):65–70.

» Chapter 14 «

Quality and Community Accountability: A View of the American Hospital Association

Thomas Granatir

There is more than a verbal tie between the words common, community, and communication. People live in a community by virtue of the things they have in common; and communication is the way in which they come to possess things in common. What people must have in common in order to form a community or society are aims, beliefs, aspirations, knowledge—a common understanding. . . .[1]

PUBLIC accountability for the quality of health care services has become one of the dominant issues of the day. For many, accountability means exposing health care organizations to external scrutiny through the publication of comparative performance information. Indeed, many initiatives are being developed to gather data from health care organizations and report information publicly. As one observer noted recently, "Report card day is here."[2]

The mechanisms for the public reporting of performance information have come to be seen as important instruments for controlling the growth of health care spending because they invoke the discipline of market forces. By giving purchasers and consumers the means to make more prudent health care decisions, the market for health care is expected to become more competitive and more efficient. Other positive effects are expected to follow, including wider participation in quality improvement by health care organizations, more discriminating accreditation decisions, and more efficient government regulation.

Special thanks to Richard J. Bogue, Deborah Bohr, Christine Izui, and Robert Sigmond.

The report card strategy conceives of accountability as something extrinsic to the organization. An organization is "held" accountable by a skeptical public. The organization is viewed as a discrete, self-contained system, and performance is judged in terms of the health care organization's "outputs." Reporting focuses on select dimensions of process or domains of outcomes. Patients recover. More enrollees have mammograms, and fewer have high-risk pregnancies. Such reports cast little light on the overall impact of health care services on public health, however.

The American Hospital Association (AHA) has developed a proposal for a more accountable health care system that focuses less on the publication of information to promote competition and more on the development of cooperative relationships with communities and their institutions to identify health priorities and work toward the achievement of major health goals. This view of accountability emphasizes the relationship of each health care organization to the larger health care system in its community. Accountability is less a matter of performance than a matter of responsibility. Accountability becomes a matter of creating a feeling of community ownership of the health care organization's strategic direction and major health goals. Organizations should align their internal quality planning and improvement programs with the community's health priorities.

COMMUNITY CARE NETWORKS: A VISION OF PUBLIC ACCOUNTABILITY

The AHA has developed a strategy for reforming the health care system to provide universal access to health care coverage, build economic discipline into the health care system, and improve the quality of care.[3,4] Focusing on public accountability for community health status, the AHA envisions a restructured delivery system that includes the establishment of community care networks organized at the local community level. This strategy has four key goals:

1. Change incentives so that resources will be managed more efficiently.
2. Integrate and coordinate services to provide a seamless continuum of care.
3. Focus on improving the health status of the community.
4. Be publicly accountable to the community.

The current system of financing health care reinforces the fragmentation of delivery. Hospital services, home health care services, and long-term care services are covered separately; services are paid for piecemeal; and providers are certified individually, each for its special function. The AHA envisions integrated delivery systems with common incentives for physicians, hospitals, and other providers that are participating in the network. Capitated payment would align the incentives of individual practitioners and providers, encouraging them to focus on disease prevention and health rather than on acute medical interventions. Integrated financing would contribute to the integration of delivery, which is less costly, more efficient, and of better quality than fragmented delivery. Furthermore, integrated financing would provide incentives to reduce duplication and to maximize the efficiency of resource utilization within the network.

Community care networks would coordinate and manage care across health care settings. The network would provide for a full continuum of health and related services; coordinate the care provided by all the constituent providers within the network; coordinate interorganizational planning; and integrate information—financial, clinical, and administrative—in a way that would allow the network to follow patients and fully manage episodes of illness.

Hospitals traditionally provide acute medical care to patients, so their traditional frame of reference for improvement is the context of medical services relevant to an acute episode of illness. As a consequence, the traditional hospital's quality concerns tend to focus on medical care rather than on health, on acute interventions rather than on prevention and wellness services, and on the patients of the hospital rather than on the community at large. Community care networks would be responsible for measuring the health status of their communities and for having mechanisms in place to improve it. Community care networks would extend their quality improvement activities outward to embrace the community and to address community health problems.

Perhaps the most important and most difficult goal of the AHA's community care network is to build a sense of community ownership of the health care system and a sense of shared purpose in the selection of community health goals. A network would demonstrate accountability to its public in two different ways. First, it would report on its performance, with a particular focus on the ways in which the network benefits the community. Second, it would include community representatives in its governing and decision-making structure, through its governing board, community advisory boards, or some other mechanism. In the network of the future, accountability would be a key organizational concern.

Accountability in this larger sense is a way to address the adverse or negative effects of competition. A managed care organization has economic incentives to reduce the duplication of services within its network, to use resources efficiently, and to channel patients to the most appropriate setting for the treatment of their health care needs. To the extent that capitated payment creates perverse incentives to provide too little care, the requirement for public accountability, with relevant measures of service use, can create countervailing incentives for the provision of services appropriate to the community's needs. Accountability mechanisms can be designed to spotlight any particular concerns in a community, such as vulnerable populations (e.g., by age or ethnic group) or diseases of special interest or urgency (e.g., acquired immunodeficiency syndrome [AIDS], or breast or prostate cancer). Thus, the focus of reporting can create targeted incentives to care for underserved populations.

More broadly, accountability to the community means taking responsibility for the impact of the organization's strategic decisions on the well-being of the community. It means taking responsibility not only for clinical performance, but also for the efficiency of the system, for helping to ensure that community resources are being efficiently and appropriately applied to priority health problems. When the forces of competition create incentives for duplication of services as each network struggles to maintain its edge, this broader notion of accountability would call for networks to consider the cost of such decisions to the community and would impel networks to collaborate in maximizing the benefits to be obtained from community resources in proportion to community need.

Community care networks, therefore, would achieve the four key goals for reforming the health care system by:

1. putting patients and communities first
2. emphasizing disease prevention and minimizing illness and disability
3. continually improving the coordination, continuity, and quality of care
4. realigning the incentives that govern everyone's behavior in making decisions about the use of health care resources

A FRAMEWORK FOR COMMUNITY ACCOUNTABILITY

The community care network would consider itself responsible for more than medical care; it would consider itself responsible for its community's health—for keeping people healthy to the extent possible through education and health promotion activities, and for following them through the continuum of care when they do get sick. The community care network would address these issues by viewing quality in the largest possible context and extending the techniques of quality management and improvement to the problems of community health.

The extension of the principles of quality improvement to community health is not a new idea. For example, the Hospital Community Benefit Standard Project embodied many of the principles of quality improvement,[5] and a multisite initiative of the Institute for Healthcare Improvement explicitly emphasizes the application of quality improvement principles to community health.[6]

The relationship between quality improvement in health care organizations and health improvement in the community becomes clear in a consideration of the core concepts of conventional quality improvement theory: leadership, team building, customer orientation, design quality and prevention, data analysis, and system thinking.[7] These concepts are broadly applicable to systems of any size and type, and thinking about how they apply to "communities" rather than to "customers" can be helpful in thinking through how best to approach community health improvement.

Leadership. Quality improvement requires a strong commitment from the organization's leaders, because the whole system must be mobilized to change. The commitment to improving quality and to putting the customer first should run throughout the organization's strategic planning. In the context of community health improvement, leadership can be asserted only through effective team building. Leaders of organizations throughout the community need jointly to demonstrate their willingness to devote themselves and their institutions to the resolution of community health problems.

Working through Teams. Effective teamwork within organizations is a key success factor for quality management. Deming identified breaking down "barriers between departments" as one of his famous 14 points for management.[8] Breaking down barriers enhances communication, coordination, and cooperation.[9] The development of partnerships is a prerequisite for bridging the leadership gap in community health improvement. Teamwork can also be a byproduct of improvement projects, because improvement is always "interdepartmental."[10] Through the development of community partnerships, shared vision and commitment can take root.

Customer-Driven Quality. In the quality planning process, who are the customers, and what are their needs are first-order questions. The ultimate measure of a health care organization's success is the level of customer acceptance that it enjoys. Everyone within the organization should be oriented to satisfying customer needs, and all of the organization's strategic planning should be directed to this goal. Similarly, defining the community and assessing its health care needs are the first steps in the development of a community health improvement plan. This decision process should be a multilayered process of interaction with the community. Just as an organization improves quality by listening to its customers, a community care network would improve health care by listening to its community.

Design Quality and Prevention. Quality improvement requires some consideration of the etiology of a problem, because it is always more desirable to address a problem at its source. A hospital may do a great job treating head injuries, for example; a communitywide program to persuade cyclists to wear helmets may in the long run be a more effective improvement strategy, however, because it will reduce the incidence of head injury in the first place.

Data Analysis. The use of data to measure progress and identify improvement opportunities is also a basic feature of the new science of quality management and improvement. Measurement and analysis, which form an objective basis for decision making, are the basic tools of improvement. Because health care organizations generally maintain data only on what they do for patients, some primary data collection would probably have to be undertaken to ensure that the information needed for community care networks is available.

System Thinking. One of the key insights of the quality improvement movement is that quality problems are problems of systems, not problems of individuals. Everyone working within an organization is part of a system that circumscribes and, to some degree, determines his or her performance. Furthermore, intraorganizational processes are not discrete or finite. The outputs of one production process may be the inputs or resources for another. Production line workers can contribute to the improvement of the production process, but only if their efforts are linked to the efforts of others who share control of the process; line workers are powerless to change the production process without the support of management and the contributions of other partners in the process.[11]

Under the old quality assurance tradition, hospital administrators saw quality fundamentally as a clinical issue and as a subsidiary function of the medical staff. This attitude persists in the current structure of organizations and in the dominance of clinical concerns in the quality and outcomes movements, which often focus on improving individual clinical decisions rather than overall organizational management.

Today, every health care organization can be seen as part of a larger health care delivery system that is, in turn, part of a social system that has enormous influence on the state of public health. Health care organizations may well rebel at the notion that they should be held accountable for the health status of their communities—measured through standard measures of population health—if they interpret such a statement to mean that they should be held *solely* accountable. Their accountability for public health should be no greater or no less than that of the production workers for the finished factory product. They can be part of a process of community health improvement, but only as part of a team, as one of several partners who share responsibilities and who control different aspects of the improvement process.

A COMMUNITY HEALTH IMPROVEMENT CYCLE

A variety of models have been suggested for the community health improvement cycle. They all reflect the same common sense cycle whose essential characteristics are assessment, action, evaluation, and adjustment.[12,13]

Assessment

Assessing Community Health Needs

The assessment process usually begins with a profile of the community and its health problems that is expected to yield a picture of community needs. Unfortunately, needs do not necessarily reveal themselves through data analysis, because they are as much matters of values, perceptions, and preferences as they are of fact. Voluntary Hospitals of America distinguishes four types of needs: (1) normative needs, based on professionally determined standards; (2) relative needs, based on comparisons with similar groups; (3) perceived needs, based on community perceptions; and (4) expressed needs, based on the community's demand for services.[14,15]

Clearly, the identification of needs is closely tied to the definition of the community, because the definition determines the dimensions of the undertaking. Although primarily a process of systematic data collection and analysis, needs assessment is not strictly a fact-finding exercise. The data gathering itself is greatly influenced by the values and preferences of the community in terms of how much priority should be given to primary data collection; of the primary data collected, how much weight should be given to expert opinion surveys versus general population surveys; of the general population surveys, how much effort should be made to tailor the surveys to the expressed needs of different population groups; and so on. No aspect of this proposed improvement process is immune from politics.

Thus, the needs assessment should not be done unilaterally. It should be accomplished through the initiating partnership.

Assessing the Health Care System

After the community's health care needs have been identified, the health care system itself must be assessed. It is essential to evaluate the capabilities, strengths, and weaknesses of the current health care delivery system, as well as other social and related services that can complement the health care system in addressing the identified health care needs. This assessment is likely to focus on the adequacy of the system, for it will be the essential resource to be balanced against measured community need. However, because the overall efficiency of resource allocation within the community is as important as access to needed services, the health care system assessment should take into account not only the system's deficiencies, but also excess capacity and overutilization of services.

Action

Developing Partnerships

Effective quality improvement within organizations depends on building teams with expertise in all the processes that affect the identified problem. Similarly, community health improvement does not happen without the support and participation of many segments of the community. This approach is necessary in part because community health problems, like all quality problems, occur within systems that have many diverse constituencies and in part because health problems have complex causes that require broad-based participation to be resolved.[16] It will always be tempting for organizations to attempt to work alone, because collaboration is so much more difficult. It will be particularly important, and particularly difficult, to involve members of the target population. Experience suggests that the rewards justify the effort, however, not just in terms of the impact of the particular intervention, but in laying the groundwork for future collaborations.[17]

Establishing Priorities

The identification of health care priorities is the most political component of the community improvement planning process. Thus, the relative strength of the organized interests participating in the work group is likely to determine these priorities. In fact, the very tendency to depend on already organized interests to make decisions about health care priorities often interjects an element of bias into the process. Each community must decide for itself how it will broaden the representativeness of the decision-making group or process.

Often, the use of some predetermined and agreed upon criteria will help make the process of deciding on priorities more rational. Such criteria may include:

- scope, or the extent of the problem
- severity, or the seriousness of impact of the problem
- feasibility, or the ability to address the problem
- cost
- importance, or the value attributed to the issue in the community

Sometimes, the criteria themselves can be prioritized. One problem may be prevalent, but have a relatively modest health or cost impact; another may occur less frequently, but have a greater impact. Every community will bring to the process its own values and preferences.

Setting Objectives

Improvement takes place in the context of goals, which should be measurable *and* attainable.

Selecting Interventions

The selection of actions will depend largely on who from the community is participating and, once again, on the careful development of agreement among all partners in the program for a particular intervention.

Evaluation

Too often, evaluation is an afterthought to the mobilization of an intervention. Just as data collection and analysis are basic management tools in a quality improvement environment, evaluation should be built into a project from the start.

Public accountability for results of the effort is an important way to ensure acceptance of the process. In addition, it provides feedback to participants, an essential component of the improvement process.

Adjustment

Quality improvement requires commitment and planning for the long term. Community health care improvement is a continuous process that evolves as community needs, resources, and understanding change. Ideally, health care organizations will link their internal and external quality improvement activities, using the process of priority setting at the community level to drive system improvements.

MODELS OF COMMUNITY HEALTH IMPROVEMENT

Community-Oriented Primary Care

The philosophical foundations of community-oriented primary care date back over two decades. In this model, clinicians, through a group practice, assume responsibility for a defined community, conduct a health care needs assessment, identify top health problems, design and implement interventions, and assess the impact of interventions.[18] Community-oriented primary care is generally physician dominated and professionally oriented, with a strong emphasis on epidemiological data and less attention to the essentially political process of developing community consensus, forming community partnerships, and achieving a sense of shared purpose among community organizations toward the solution of health problems. In community-oriented primary care, the group practice assumes responsibility for "the needs of the community in which it resides and for its own impact on the community."[19]

Community Health Intervention Partnerships (CHIPs)

A national demonstration program funded by the Robert Wood Johnson Foundation and the United Hospital Fund, the CHIPs program is currently operational at five sites. The CHIPs process includes the following elements:

- partnership building among hospitals, other health care providers, and community-based organizations. The project works to develop partnerships at several levels, including full working partner, steering partner, linking partner, and informing partner.
- collaborative needs assessment. In a participatory, democratic process, the community health care needs assessment project involves searching for epidemiological data and gathering more subjective information from the community, with a focus on strategies for understanding the etiology of problems and identifying strategies for change.
- priority setting and action planning. Community partners jointly develop plans to meet priority health care needs.

The focus of intervention in CHIPs ranges from health service issues (e.g., substance abuse, human immunodeficiency virus [HIV] infection, teen pregnancy, heart disease, mental illness) to public health issues (e.g., violence) to social issues (e.g., resource coordination, multiculturalism). The distinguishing characteristic of this model, however, is its emphasis on partnerships—on organizing within the community, empowering community organizations, and building acceptance for community-based interventions.[20]

Hospital Association of Pennsylvania "Planting the Seeds for Good Health"

The Hospital Association of Pennsylvania has developed a model for assessing and improving the health status of communities in Pennsylvania.[21] The model suggests a process in phases, including:

- the construction of a county health profile, including demographics, health care resources, health status indicators, and small area variation analyses
- a structured survey of the health behaviors of individuals in the community; an inventory of all health-related organizations, agencies, providers, and referral organizations in the health and social service community; and, finally, an economic impact statement mapping the flow of financial resources in and around the community's health care system
- the identification of priorities for action on community health care needs through focus groups with consumers, business leaders, educators, and health care leaders
- the development of a community action plan; finding of project partners

This model is a highly structured and essentially data-driven model with a strong focus on community leadership. Extensive analysis of secondary and primary data is used to profile community health and resources; then structured focus groups are used to identify, establish priorities of, and plan to address community health care needs.

Hospital Community Benefits Standards Projects

Established at New York University and funded by the W. K. Kellogg Foundation, the Hospital Community Benefits Standards Projects developed a set of standards to help hospitals in their efforts to meet their community benefit objectives. The standards require a hospital to set objectives, develop plans, and carry out activities that aim to improve the health status of community members, provide services to underserved groups, and reduce the growth of health care costs in a community. The Catholic Health Association and the Voluntary Hospitals of America have developed similar approaches for community benefit. The standards provide a template for hospitals to affirm their commitment and assert their leadership in community health improvement projects. Although the standards address a hospital's participation in projects that it does not lead, the overall emphasis of the standards is on the leadership that the hospital demonstrates in catalyzing the community to action.[22,23]

Planned Approach to Community Health (PATCH)

A comprehensive health care planning program, PATCH is designed to build analytical capacity and effectiveness of local health care planning agencies. The PATCH process consists of diagnosing and assessing health problems, identifying the risk factors associated with those problems, and intervening to attack the causes of those risk factors. As a prevention program, PATCH emphasizes the systematic movement "upstream" from health problems to root causes.[24]

Communitywide Health Improvement Learning Cooperative

The Institute for Healthcare Improvement and GOAL/QPC are sponsoring projects in nine locations across North America to test the application of continuous quality improvement methods to problems of community health improvement.[25]

MAJOR CHALLENGES

Defining Community

Traditionally, physicians and hospitals define their "communities" in terms of their patient populations—those individuals using the practice or admitted to the institution. In this sense, community can be defined only historically. On the other hand, a health maintenance organization ordinarily defines its community in terms of its predefined enrolled population. For these organizations, community is defined not just in terms of users of the system, but in terms of potential users as well. Communities can also be defined in terms of specific problems (e.g., individuals with certain diseases) or in terms of specific interests (e.g., ethnic or generational issues). Most broadly, community is defined by a perceived sense of commonality, whether social, cultural, economic, or moral. Most practically, community is defined in terms of a relatively circumscribed political or geographical service area within which there is a sense of interdependence and belonging.[26]

It is not easy to define community, but the definition should be an explicit part of the process of finding partners for health care improvement projects. The definition of community affects every aspect of the process—what needs are identified, who will be involved in establishing improvement priorities, and who is going to benefit—so it must be done explicitly. Furthermore, it will affect the scope of the interventions, so it must be done thoughtfully.

Finding the Data

A large number of secondary data sources are available to perform the needs assessment: census data; provider billing data for Medicare and, in many states, non-Medicare admissions; cancer and other disease registries; and vital health records. Useful data can be obtained through local police, as well as through state transportation and labor departments. The availability of many of these data in electronic formats and nearly universal coverage makes them inexpensive and attractive. On the other hand, the data sets may be difficult to link together meaningfully, it may not be possible to sort the data according to local definitions of community, and the data may not be quite focused on the issues of concern to the community. Furthermore, because of the size of the data sets, the data may not be timely.[27,28] Nevertheless, secondary data can be useful for performing epidemiological studies, for calculating service statistics for providers, for calculating some population measures, and for developing the inventory of resources. Although helpful, secondary data sources will probably not prove sufficient.

Primary data collection of various types—general surveys or surveys of select subpopulations, surveys or interviews of experts, focus groups, and public meetings—can be effective and versatile ways to approach the needs assessment. Each method has its advantages and disadvantages. Surveys can be highly valid and reliable, but are often costly. On the other hand, public meetings may be inexpensive; may open up the process to the community at large, including unorganized populations; and can create a sense of acceptance for the process. Public meetings are likely to yield biased, or at best unrepresentative, information, however. Data collection for needs assessment can be eclectic. The Viroqua, Wisconsin CHIPs facility used survey, focus groups, and a review of secondary data. Six different survey instruments were used in all, each tailored to one of six target population groups, reflecting the importance attached to the recognition of cultural diversity in this project.[29]

Recognizing the Importance of Values

The differences in overall concept of prevailing community health care improvement models are really quite small, but their differences in emphasis have important practical implications. For some, the emphasis is on assessment and analysis of health care needs and options. They emphasize the data, seeing the task of community health care improvement as essentially a technical exercise. The data will reveal what the community needs. The experts with knowledge of health and disease, planning, and program evaluation will be best suited to suggest "treatment options." Priorities may be driven more by feasibility and cost than by values and preferences. Partnerships are formed after needs are assessed, primarily as a means to implement interventions.

For others, the emphasis is on identifying priorities, developing partnerships, and building teams. They emphasize personal or organizational factors. They believe that it is more important to listen to people than to attend to the data and that values carry as much weight as facts, particularly when it comes to selecting priorities for health care action. They place a priority on partnerships as a means of including the disenfranchised, establishing a feeling of consensus within the community, and making a commitment to change.

Values matter, particularly in enterprises like efforts to improve the quality of health care. Although heavily dependent on data and analysis, and despite their roots in the application of statistical techniques to production processes, the quality management philosophies expounded by Deming and Juran are essentially humanistic. According to these philosophies, the quality of the organization is a function of such factors as leadership, teamwork, and personal empowerment. Success depends on how well the organization has listened to its customers.

CONCLUSION

The science of quality improvement has been revolutionized in the last decade by the application of epidemiological methods to quality oversight; this approach has established a perception of poor quality as a disease of the health care system to be screened for and treated. Paradoxically, as community care networks assume responsibility for improving community health, they will be adapting the methods of continuous quality improvement to the epidemiological problems of public health.

There is no doubt that the reform of the health care system will bring much wider public exposure of performance information about health care organizations. This public exposure is being labeled a mechanism for enforcing accountability, as if accountability were something extrinsic to the organization. The development of systems for measuring and reporting on performance would be a marvelous accomplishment, but measuring and reporting alone will not satisfy the need for a more accountable health care system.

Two years ago, the AHA proposed a vision for reforming the health care delivery system through organized networks of care that assume responsibility for improving the health status of their communities. It is already happening in dozens of communities across the United States, as health care organizations demonstrate their commitment to improving community health through a merging of public health planning and quality improvement traditions in locally organized partnerships.

NOTES

1. J. Dewey, *Democracy and Education* (New York: Macmillan and Company, 1916), 4.
2. D. O'Leary, The Measurement Mandate: Report Card Day Is Coming, *Joint Commission Journal on Quality Improvement* 19, no. 11 (1993):487–500.
3. American Hospital Association, *Transforming Health Care Delivery: Toward Community Care Networks* (Chicago: 1993).
4. American Hospital Association, *National Health Care Reform: Refining and Advancing the Vision* (Chicago: 1992).
5. American Hospital Association, *Transforming Health Care Delivery*.
6. R. Veatch and B. Bader, Measuring and Improving Community Health, *Quality Letter for Healthcare Leaders* 6, no. 5 (1994):2–21.
7. The term *system thinking* has taken on a special meaning through the work of Peter Senge, author of *The Fifth Discipline*. I am using the term colloquially here.
8. M. Walton, *The Deming Management Method* (New York: Perigee Books, 1986).
9. Joint Commission on Accreditation of Healthcare Organizations, *Using Quality Improvement Tools in a Health Care Setting* (Oak Brook Terrace, Ill.: 1992).
10. J.M. Juran, *Juran on Leadership for Quality: An Executive Handbook* (New York: Free Press, 1989).
11. Walton, *Deming Management*.

12. Voluntary Hospitals of America, *Community Health Assessment: A Process for Positive Change* (Irving, Tex.: 1993).
13. Hospital Research and Educational Trust, *Background and Resources for a Community Health Status Focus* (Chicago: 1993).
14. Voluntary Hospitals of America, *Community Health Assessment*.
15. United Way of America, *Needs Assessment: The State of the Art. A Guide for Planners, Managers, and Funders of Health and Human Care Services* (Alexandria, Va.: 1982).
16. Voluntary Hospitals of America, *Community Health Assessment*.
17. J. Trocchio, The Hows and Whys of Conducting a Community Needs Assessment, *Trustee Magazine*, March 1994.
18. P.A. Nutting, *Community-Oriented Primary Care: From Principle to Practice* (Washington, D.C.: Health Resources and Services Administration, HRSA Pub. No. HRS-A-PE-86-1, 1987).
19. L.H. Berman, Defining and Characterizing the Community: The Use of Survey Research Techniques and Tools in COPC, in *Community-Oriented Primary Care: From Principle to Practice*, ed. P.A. Nutting (Washington, D.C.: Health Resources and Services Administration, HRSA Pub. No. HRS-A-PE-86-1, 1987), 134–151.
20. Hospital Research and Educational Trust, *Three Year Implementation and Replication Plan for Community Health Intervention Partnerships* (Proposal to the Robert Wood Johnson Foundation, 1994).
21. Hospital Association of Pennsylvania, *Community . . . Planting the Seeds for Good Health. A Guide for Assessing and Improving Health Status*, 1993.
22. Hospital Research and Educational Trust, *Background and Resources*.
23. T. Hudson, Hospitals Strive To Provide Communities with Benefits, *Hospitals*, July 5, 1992.
24. Centers for Disease Control, *Planned Approach to Community Health (PATCH): Program Descriptions* (Atlanta, Ga.: 1992).
25. Veatch and Bader, Measuring and Improving Community Health.
26. Hospital Research and Educational Trust, *Background and Resources*.
27. A. Trachtenberg et al., Building an Integrated Community Health Information Base, in *Community-Oriented Primary Care: From Principle to Practice*, ed. P.A. Nutting (Washington, D.C.: Health Resources and Services Administration, HRSA Pub. No. HRS-A-PE-86-1, 1987), 109–125.
28. United Way of America, *Needs Assessment*.
29. Hospital Research and Educational Trust, *Three Year Implementation*.

» Chapter 15 «

From Theory to Practice: Managing Quality under Cost Constraints

Timothy C. McKee and Timothy J. Ward

THE availability of resources is important in defining the appropriateness of health care services. If resources are not constrained, then the appropriateness of health care services depends solely on its quality, that is, the ability of the clinical intervention to affect health status. If resources are constrained, then the definition of appropriateness is modified to take into account the relationship between quality and the cost of the health care services. In both cases, the debate focuses on the appropriateness of choices made by the patient and the provider. In the first case, cost is not a consideration; in the second case, cost is an integral part of the appropriateness judgment.

In both cases, it is difficult to determine the validity of the choices because of the lack of empirical evidence to clarify the link between the outcomes and processes of care. Given unconstrained resources, quality is perceived as a conceptual absolute that would be defined by the weight of good empirical evidence—if it were available. Therefore, the basis for the debate in this case is the lack of knowledge about the outcome-process relationship and would presumably dissipate with the collection of better empirical data. Given constrained resources, however, choice is restricted not only by a lack of knowledge, but also by resource limitations. With constrained resources, the availability of good empirical evidence would only sharpen and intensify the debate over priorities, not remove it.

It is important to clarify the use of the terms *costs* and *resources* in this chapter. Cost is a dollar measurement that results from numerous decisions, both clinical and administrative, about the consumption of individual resources. For example, individual provider treatment decisions determine the volume of health care services produced by a plan. When the plan owns its capacity, then administrative decisions on sizing, equipping, and staffing determine fixed cost. Because most costs are fixed, these capacity decisions largely determine unit costs. Total cost is a function of both clinical and administrative decisions. It is necessary to uncover the relationship between clinical and administrative resource decisions in order to manage quality and cost.

THEORETICAL MODEL OF APPROPRIATENESS

Shifting the definition of appropriateness from a singular and absolute judgment based solely on quality to a relative judgment based on quality and cost introduces two problems. The first is actually defining the appropriateness of care when it is based on relative judgments. The second is managing the complex issues that emerge when the definition of appropriateness is also relative across different communities or plans.

Defining Appropriateness

For this discussion, it is assumed that the individual faces no financial barriers to access. Costs are considered constrained for the larger group, however, whether it be the provider, the employer, or the region. Therefore, because it is not possible to do everything for everyone, choices will have to be made.

Under constrained resources, appropriateness is defined by best clinical practice. Given alternate choices of treatment that afford the same outcome for the patient, best clinical practice is the treatment that requires the lowest resource consumption. This definition of best clinical practice is based on the assumptions that the treatment choice will not affect the outcome for the patient, but that cost will be a major factor in judging appropriateness. For example, if a region has a Caesarean delivery rate that varies from 12 to 35 percent and Caesarean deliveries are more expensive than vaginal deliveries, then best clinical practice will be defined as the lowest Caesarean delivery rate that provides similar outcomes for the mothers and babies.

Treatment choices defined by best clinical practice do not imply that the care provided at other locations is poor in quality. All the Caesarean deliveries at a location with a 35 percent rate could be skillfully executed and result in excellent outcomes for mothers and babies, but this location would not be applying best clinical practice. A location with a 12 percent Caesarean delivery rate and similar excellent outcomes to mothers and babies would consume fewer resources, and this location would define the best clinical practice frontier. Changes in treatment choice based on a definition of best clinical practice do not represent a decline in quality, but rather an opportunity to manage resource consumption.

The definition of best clinical practice is an iterative and ongoing process of debate. The debate centers on the persuasiveness of empirical evidence regarding patient outcomes in relation to the resources consumed by the treatment. Under constrained resources, the debate is not focused solely on quality of care but on the relationship of quality to cost.

Defining Best Clinical Practice across Groups

Different providers, employer groups, or other subpopulations may define appropriateness of care differently. If appropriateness is a function of cost and quality,

$$\text{Appropriateness} = f(\text{Cost}, \text{Quality})$$

the definition of appropriateness depends on the definition of the variables in the algorithm. Such differences arise naturally from the theoretical construct of appropriateness.

Appropriateness: Individual or Population?

The focus for the definition of appropriateness can be an individual or a population. In the Caesarean delivery example, best clinical practice was defined for the individual based on the lowest cost that produced identical patient outcomes. This approach implies that an absolute judgment of appropriateness can be based on minimizing costs while maintaining quality of care for the individual. The judgment of appropriateness may be very different if the health of the population rather than that of the individual were the focus. With best clinical practice defined in terms of a population, judgment hinges on whether it is appropriate to trade individual quality of care for improvements in the health status of the larger population.

Eddy has suggested that it may be appropriate from a quality perspective and necessary from a cost perspective to provide less than optimal quality for a single patient if the population as a whole benefits.[1,2] For example, radiologists may have a choice between two contrast agents, ionic and nonionic, when performing radiographic procedures. Both are considered safe and effective by the medical community and are approved by the Food and Drug Administration. The nonionic agents have fewer side effects, but cost 10 to 20 times as much. If the ionic agents are used exclusively for the 42,000 procedures performed in a large managed care plan, the number of mild reactions can be expected to increase each year by 1,000; the number of moderate reactions, by 100; and the number of severe reactions, by 40. Using the ionic agents would free $3.5 million each year for other purposes. If the money were spent on increased breast cancer screening, it could prevent 35 deaths. If it were spent on cervical cancer screening, such funds could prevent 100 deaths. If the $3.5 million were spent on cholesterol treatment, it could prevent 13 sudden deaths; 105 heart attacks; and 250 cases of coronary insufficiency, angina, and other coronary artery disease events. Eddy concluded that the quality of care provided by the plan would be higher if the lower cost agent were used and the savings were applied to other interventions. Clearly, the appropriateness of provider and patient choices will be judged differently, depending on the view taken when defining best clinical practice.

Cost: How Constrained Are Resources and for Whom?

Costs are not borne evenly by all individuals in the health care delivery setting, much less in the health care system. A patient with no responsibility for co-payment may demand far more diagnostic tests than either a capitated provider or a patient with a large deductible. The attending physician may recommend additional tests in part to reduce personal liability associated with malpractice. Both the patient and the physician may differ with the director of the pharmacy who, in order to meet budget constraints, has restricted access to an expensive drug. Because the nature and amount of the cost that they incur will influence their views on priorities and on opportunity costs, it is worth noting which costs affect whom and by how much.

The severity of the cost constraint will also influence the way that affected parties respond to the quality-cost trade-off. This is particularly true when the severity highlights the opportunity costs of different choices. Both patients and medical professionals can be expected to tolerate unavoidable inconveniences brought on by clear and unavoidable constraints. They are less likely to be tolerant if constraints are believed to jeopardize quality, however. With different opportunity costs, and depending on the severity of the constraints, different players will define best clinical practice differently.

Quality: Scope Is Everything!

Donabedian defined three major components of quality.[3] One component is the application of the science and technology of medicine, and of the other health sciences, to the management of a personal health problem. The second component is the management of the social and psychological interaction between client and practitioner. The third component, which Donabedian called amenities, includes such features as pleasant and restful waiting rooms, a comfortably warm examination room, clean sheets, a telephone by the bed, and good food.

The definition of best clinical practice may depend on whether the definition of quality has a narrow or a broad scope. For example, when evaluating the appropriateness of differences in Caesarean delivery rates, some providers may weight patient satisfaction heavily, while others may give significant weight only to the technical component. Therefore, different Caesarean deliveries could be deemed appropriate by different providers.

Dealing with the Relativity Dilemma of Best Clinical Practice

Considering appropriateness to be relative to both cost and quality removes some of the stability from the clinical decision-making process that a more absolute definition of quality provided. The reality of cost constraints requires providers and patients to explore the "shaky ground" of choice and trade-offs. Rather than allowing the debate to remain a comfortable distance away from the treatment process and within the realm of clinical and health services researchers, the reality of constrained costs forces individuals to deal with the dilemma of choice.

Although the definition of appropriateness may be less stable under constrained costs, assessment of technical quality still provides a fairly firm floor from which to judge treatment decisions. Variations in best clinical practice that result from differences in technical quality should not be tolerated. The judgment of the "goodness" of technical quality remains dependent on the appropriate application of medical science and technology in affecting health status.[4] In the Eddy example, technical quality is still firmly defined, although trade-offs are permitted between the individual and population health status. Likewise, the rate of Caesarean deliveries may vary, depending on best clinical practice, but the technical outcome of the Caesarean delivery for the mother and the baby should be more absolute. The challenge is to guide interested parties through a rational process of defining appropriateness above the absolute floor established by technical quality. The beliefs, values, and priorities held by the individuals involved will influence the process. Some participants may try to minimize best clinical practice differences across groups while others may be more comfortable with these differences.

UNIFIED MODEL

To be useful in defining best clinical practice, the unified model must help with two difficult tasks. It must organize the diverse "players" into a systematic structure in order to promote the dialogue required for informed choice, and it must direct action. The model should be judged on its usefulness in accomplishing these two tasks.

The unified model has three main components (Figure 15–1), which must work together in order to define best clinical practice. These components are information feedback, clinical enterprise, and capacity engineering. The information feedback component provides the structure necessary to interpret comparative data on cost, quality, and access performance. This component helps guide and focus data interpretation. Information by itself does not change practice patterns, however; it must be incorporated into the provider's treatment decision processes.

The clinical enterprise component builds on information obtained in the information feedback function to effect changes in provider practices. The clinical enterprise function encompasses all professional activities concerned with defining best clinical practice.[5] Changes in provider behavior do not translate effectively into cost savings unless capacity can respond appropriately to changes in clinical volume, however.

The capacity engineering component is critical to managing best clinical practice in two ways. First, it promotes valid analyses by identifying achievable savings, rather than average savings, as a result of clinical changes. Second, the capacity engineering tools guide the implementation decisions required to realize the savings promised in the analyses. The model is dynamic, and its effectiveness depends on each component performing its function.

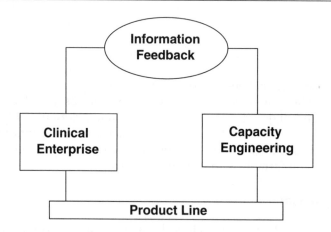

Figure 15–1 Unified Model of Health Care Delivery

Finally, a product line focus is one of the most important features of the unified model. Product lines are necessary for linking input costs with quality outcomes, thus providing a focus for clinical and administrative management actions.

Information Feedback Function

Although it would seem that the data are more important than the manner in which the data are structured or organized, structure and data are equally important characteristics in determining the effectiveness of the information feedback function. There are two reasons. First, structure is a conceptual decision that is within the control of the providers or administrators, while data issues are usually driven by issues beyond their control. There is significant value in beginning with what is under their control. Second, structure provides a guide for the interpretation of the data, and proper interpretation is at least as important as the data itself.

The structure of the information feedback function integrates its four components sequentially. The first component uses a productivity framework to define performance. The second component identifies the separate productivity functions within a population-based model of health services delivery. The third component describes the relationship of cost, quality, and access performance parameters to the productivity framework. The fourth component builds a matrix that captures the relationships between the productivity functions and performance parameters. The matrix is the foundation for structuring the relationships among the data elements used to improve health care system performance.

Productivity Relationship

The whole point of struggling to identify best clinical practice is to improve the health care system performance. In industrial engineering and other management sciences, output divided by input equals productivity. In the health care system, performance can be equated with productivity.

$$\text{Performance} = \text{Productivity} = \text{Output}/\text{Input}$$

The struggle for improved health care system performance is based on active management of the quality-cost relationship in striving to define best clinical practice. Therefore, the concept of productivity, in which the quality of clinical care is the output and the cost of the resources is the input, is the source of best clinical practice. It makes sense that the information both to monitor and to improve health care system performance must be presented in a productivity framework. Accordingly, the structure of the information feedback component must be based on productivity relationships.

Epidemiological Foundation

Although productivity relationships provide rigor in defining performance and focus for the information feedback function, how can the productivity relationships inherent in a patient treatment process be identified? In addition, how is it possible to ensure that the relationships identified are comprehensive in describing the patient care process? The answer to these questions can be found in an early work by Lee and Jones, who in 1932 produced the first study that related resources to the health care needs of the U.S. population.[6] They used an epidemiological model that remains current today and is the epidemiological foundation for the unified model.

The epidemiological model is elegant in its simplicity. As shown in Figure 15–2, the volume of resources required to support a population's health care needs depends on the prevalence of illness in the population, the volume of health services consumed in treating the illness, and the labor and capital used to produce those services. Therefore, the key to managing health care costs for a population depends on managing the relationship among these elements.

The relationship among the elements of the epidemiological model can be seen more clearly in Figure 15–3. Health promotion activities focus on the relationship between a population and the prevalence of disease within the population. The goal of health promotion programs is to reduce illness in the population. Analysis of utilization patterns explores the relationship between patient disease and the volume and type of health services used to treat the disease. There is a rich body of literature analyzing variation in utilization patterns across populations and regions.[7] Finally, the treatment pattern characterizes the relationship between the health services used to treat the patient and the resources consumed in producing those health services.

The power of the epidemiological framework is the simplicity with which it identifies the opportunities to manage costs. As shown in Figure

---

Population ⇨ Illness ⇨ Health Services ⇨ Resources

Figure 15–2 Epidemiological Foundation for Health Care Delivery Model

Figure 15-3 Process of Managing Volume

15-3, managing health care costs for a population depends on managing health promotion activities, utilization patterns, and treatment patterns. In order to improve performance in these areas, however, it is necessary to identify the productivity functions that underlie these activities.

As shown in Figure 15-4, the output/input or productivity relationships make it possible to identify the individual contributions of disease and provider practice decisions to differences in the total volume of services consumed by different populations. It is important in any monitoring system to distinguish those causes of variation that are exogenous or outside of the health care system (e.g., disease) from causes that are endogenous or within the control of the system (e.g., provider treatment choice).

Unfortunately, most data sources provide very little insight into the prevalence of disease across populations. Therefore, as shown in Figure 15-5, there is only a single productivity relationship that compares the health status of a population with the volume of episodes consumed by that population. In this unified model, the rate at which populations consume health services is called utilization productivity.

The confounding of disease prevalence and physician practice patterns mandates caution in interpreting the significance of differences in utilization productivity across populations. Although it may be that there are fundamental differences in the disease epidemiology, it may also be that

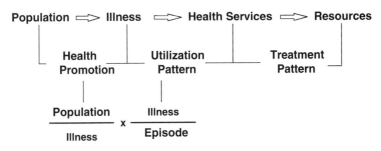

Figure 15-4 Process of Managing Volume

278 THE EPIDEMIOLOGY OF QUALITY

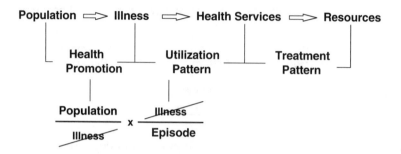

Figure 15–5 Process of Managing Volume

health services are being consumed at a different rate for a similar disease incidence. It is necessary to be equally cautious in directing management actions in response to utilization productivity differences. A choice must be made between committing limited resources to health promotion efforts to reduce disease incidence and building control systems to influence provider behavior.

Finally, embodied in the epidemiological model's treatment pattern are two additional productivity relationships (Figure 15–6). These two productivity relationships result from an intermediate step in the production of health care episodes. Labor and capital inputs are combined in various hospital departments to produce products that are intermediate in the production of patient episodes. These intermediate products are then combined within clinical departments to produce the completed episode. For example, laboratory tests, drugs, and intensive care unit time are all intermediate products that are produced by the hospital and then combined by providers to complete episodes of care. Hospital department pro-

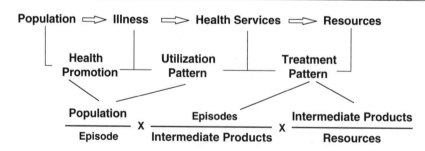

Figure 15–6 Process of Managing Volume

ductivity relates the input of labor and capital to the intermediate product outputs. Clinical department productivity accepts the intermediate products as inputs in producing patient episodes. All three productivity relationships are shown in Figure 15–7.

In Figure 15–8, simple math demonstrates that the productivity relationships reduce to per capita expenditures. Therefore, management of these three productivity relationships will manage care for a population. Accordingly, these productivity relationships become the focus for the information feedback function.

Cost, Quality, and Access Parameters

Donabedian defined cost, quality, and access as the major parameters for measuring performance in health care delivery.[8] Thus, to complete the basic evaluation structure of the unified model, it is necessary to relate these parameters to the productivity relationships. Quality, being patient-centered, is a predominant characteristic of the output, while cost is the predominant characteristic of the input. Because the quality/cost relationship is inherent in the output/input relationship, the quality/cost trade-off is reflected in all productivity decisions. This trade-off is interwoven in the decision process, whether producing intermediate products, episodes of care, or the health of a population.

Access warrants a somewhat different treatment. Access is very broadly defined to include availability, accessibility, accommodation, affordability, and acceptability to patients and providers,[9] as well as continuity and coordination of care. Rather than an attribute of the productivity relationship itself, access describes the impact of the larger system on the consumption of intermediate products and episodes of care, and on the change in a population's health status. Of course, access affects costs and quality, but it is not a part of the productivity relationship itself.

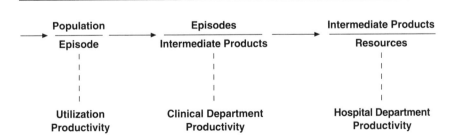

Figure 15–7 Productivity Relationships

280   The Epidemiology of Quality

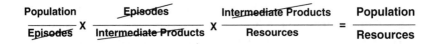

Figure 15–8 Mathematical Reduction of Productivity Relationships

---

The relationship of cost, quality, and access to productivity is presented in Figure 15–9. Quality is shown as a characteristic of the output; cost is a characteristic of the input; and access affects the consumption of the product or service. Clearly, the cost, quality, and access parameters can be integrated with and fully describe a productivity relationship.

Matrix for Evaluation

Rigorous definitions of both the productivity relationships and the performance parameters result in a structure that has two major advantages for producing effective information feedback. First, such rigor enables a mass of data to be appropriately organized. Second, a meaningful organization of the data facilitates a valid interpretation of the data.

Figure 15–10 illustrates a matrix that integrates the epidemiologically based productivity relationships with cost, quality, and access performance parameters. The productivity relationships form the rows with the cost, quality, and access parameters displayed along the columns. Because both the productivity functions and performance parameters are defined comprehensively, all relevant health care data will fit into one of the matrix's nine cells.

Experience has shown this matrix to be very useful in organizing health care data, as it provides a filing system to categorize available data and to identify missing data. For example, the matrix is useful in evaluating the

---

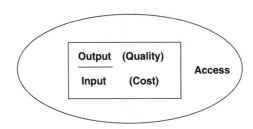

Figure 15–9 Relationship of Cost, Quality, and Access to Productivity

	COST	QUALITY	ACCESS
UTILIZATION PRODUCTIVITY			
CLINICAL DEPARTMENT PRODUCTIVITY			
HOSPITAL DEPARTMENT PRODUCTIVITY			

Figure 15–10 Performance Matrix

---

impact of an automated data system on improved productivity in relation to the cost of data collection. Because the matrix is comprehensive, it also provides a structure to prioritize system requirements.

In addition, the matrix helps in the difficult task of interpreting the data, once they have been collected, by forcing a comprehensive examination of cost, quality, and access measures. For example, in the area of women's health, Caesarean delivery rates are often used as a proxy to measure the cost reduction potential among health care plans. The matrix structure illustrates the difficulty of interpreting the significance of isolated Caesarean delivery rates without the accompanying quality data on the outcomes for the mother and baby, however. In order to complete each row of the matrix, it is necessary to predict and evaluate cost, quality, and access data.

The lines between the rows signify that each productivity relationship should be interpreted separately. For example, the production of intermediate products should be considered separately from the way in which these products are clinically combined to produce a patient episode. Therefore, the matrix suggests a sequential order in which to examine performance. Describing the productivity relationships separately in the matrix is also a reminder that the full set of epidemiological productivity relationships should be investigated.

Within the matrix, measures must be chosen to populate the matrix cells. The need to choose among measures introduces the richness and rigor of the structure-process-outcome paradigm.[10] Structural measures describe the physical components of a system (e.g., provider credential committees). Process measures focus on the content of care (e.g., immunization rates). Outcome measures describe the impact of care on the patient (e.g., health status). Although Donabedian established this paradigm to describe measures of quality, the paradigm works well in discussions of access and cost parameters. For example, in measuring access to primary care, the number of primary care providers per 1,000 population may serve as a structural measure; patient waiting times, as a process measure; and percentage of outpatient or ambulatory visits seen through the emergency department, as an outcome measure. No particular category of measures is superior to the others, but it is important that the type of measure selected be suitable to its task.[11] Furthermore, because each category of measures may be interpreted somewhat differently, a knowledgeable selection is important in the final definition of best clinical practice.

Although the matrix can help in the longer term by identifying and prioritizing information system requirements, the matrix cannot solve the short-term problem of missing data. The matrix provides a constant reminder of the missing data and, therefore, cautions an interpreter of performance to proceed carefully.

A matrix with all its cells populated appears in Figure 15–11. The measures selected in this exhibit describe performance at a summary system level. Cost, quality, and access data are provided for each of the productivity relationships.

Product Lines

General data provide useful insights into general performance trends and may identify particular problem areas. Because the information is not clinically specific, it is not very useful in directing specific action, however. Without specific data, it is difficult to make the patient-provider link that enables practitioners to define and implement best clinical practice. The answer is to organize the information feedback system (i.e., the matrix) around product lines. Guidelines for establishing the product lines include the following:

- A product line should be population-focused, disease-based, and clinically relevant.
- A product line should not focus on a specialty or a procedure. For example, coronary artery disease may be a suitable product line, while cardiology or coronary artery bypass graft surgery is not. Although cardiologists, among others, may treat patients and the coronary artery bypass graft procedure is one form of treatment, coronary artery disease is the illness foundation of the product line focus. This

Managing Quality under Cost Constraints 283

	COST	QUALITY	ACCESS
UTILIZATION PRODUCTIVITY	Admissions per 1,000 beneficiaries	Patient satisfaction with quality	Patient satisfaction with access
CLINICAL DEPARTMENT PRODUCTIVITY	Average length of stay	Infection rate	Waiting time for appointment
HOSPITAL DEPARTMENT PRODUCTIVITY	Cost of prescriptions	Accuracy of laboratory tests	Turn around time for X-ray results

Figure 15–11 Performance Matrix: General

---

approach is consistent with the epidemiological foundation of the monitoring system.
- The link between inputs and outputs should be clear in a product line. For example, obstetrics is a good product line choice, because the staff and facilities are usually easy to identify. Resource modifications can be linked to changes in clinical practice, making it easier to direct management action to reduce costs.
- Because product lines should be worth the time and effort required to collect and evaluate the performance data, their volume, cost, and variability should be high. The degree of variability is important because it suggests practice differences that may benefit from best clinical practice definition.

Figure 15–12 shows a matrix that could be used to gauge the performance of an obstetrical product line. Each cell of the matrix should be populated with multiple performance measures; one or two example measures are shown in each cell in Figure 15–12.

The first row of the matrix identifies important utilization patterns. These measures are a function of the population/episode (output/input) relationship defined previously. Patients admitted to the hospital consume a large amount of obstetrical resources. While a short inpatient stay is likely for patients after a vaginal birth, two additional practices use a significant portion of obstetrical unit capacity. These two practices are the rate of admission for nondelivered obstetrical patients (i.e., patients who are admitted to the hospital for observation, testing, or some procedure,

	COST	QUALITY	ACCESS
UTILIZATION PRODUCTIVITY	Non-delivered admissions rate Caesarean section rate	Percentage low birth weight babies	Percentage first prenatal visit in 1st trimester
CLINICAL DEPARTMENT PRODUCTIVITY	Average length of stay for DRG 373	Neonatal mortality rate	Availability of 3rd level facility
HOSPITAL DEPARTMENT PRODUCTIVITY	Average cost per bed day	Accuracy of diagnostic tests	Average waiting time for sonogram

Figure 15–12 Performance Matrix: Obstetrical Product Line

---

but do not deliver an infant during this admission) and the Caesarean delivery rate.

Clearly, if the rate of admission for these two practices falls below some lower limit, the quality of care provided to new mothers and their babies will suffer. Prior analysis suggests that a great deal of variation occurs above these lower limits, however. Best clinical practice should identify the desired target values or practice policies.

The second row of the matrix identifies important clinical treatment patterns. These measures are a function of the episodes over intermediate products (output/input) relationship defined previously. The average length of stay for patients following uncomplicated vaginal birth would fit into the cost/clinical treatment pattern cell of the matrix. As noted earlier, best clinical practice will be defined as the shortest length of stay that does not adversely affect the quality of care for new mothers or their infants. Additional cost/treatment measures may include the average length of stay for patients following Caesarean delivery, the percentage of patients given epidural anesthesia, and the average number of nursing hours per birth.

As an episode-based treatment pattern performance measure, the neonatal mortality rate should focus on those neonatal deaths that are preventable or that result from the labor and delivery process. Care should be used to distinguish birth process–related deaths from those neonatal deaths that result from congenital defects (e.g., anencephaly) and are not associated with the medical care received. Other important treatment pattern/quality measures include the number of hospital readmissions, the number of unplanned returns to surgery, and the incidence of infections and excessive blood loss. The rate of episiotomy and the severe (i.e., third- and fourth-degree) tear rate among episiotomy and nonepisiotomy patients are interesting treatment pattern performance measures as well.

The third row of the matrix identifies important hospital department patterns. These performance measures are a function of the intermediate products over costs or resources consumed. In addition to the average cost of providing 1 day of inpatient care on the obstetrical unit, quality measures at the hospital department level may include measures of the accuracy of diagnostic tests.

With the product line definitions completed, the monitoring system resembles the layout presented in Figure 15–13. Following general summary measures of performance, the product lines form chapters in a report that is enriched periodically with longitudinal data. These structured data reports form the first component of the unified model of defining and implementing best clinical practice.

Obstetrical Case Study: Information Feedback

The following case study illustrates the potential impact on costs and quality of defining and implementing best clinical practice for obstetrics in a large metropolitan area. It is based on actual data from a U.S. metropolitan area. Although the cost savings are estimates, they reflect what could possibly be achieved with this model.

> A large managed care plan has enrolled 94,141 females between the ages of 15 and 44. The plan is organized as a staff model health maintenance organization and has a significant investment in physical facilities that have been acquired through the years. The beneficiaries living in the area are served by the plan's six hospitals, as well as by a number of outpatient clinics.
>
> Because of increased competition with other health care providers in the area, the plan faces increasing pressure to reduce its costs. Maintaining a high standard for the quality of care provided to its enrollees is important both in terms of the plan's corporate culture and its future marketing success, and the plan administrators have decided to implement the unified model. They have chosen obstetrics as the first product line. Obstetrics, which accounts for 30 percent of the plan's total inpatient days, was chosen because of its high volume and high cost.
>
> The information feedback system has revealed wide variations in the way that the plan provides obstetrical care. Indicators such

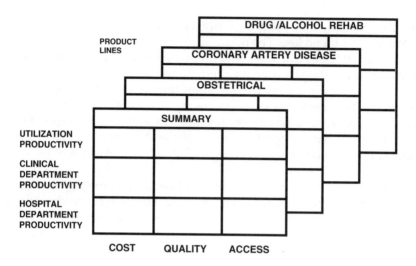

Figure 15-13 Performance Matrix Reports

as nondelivery admission rates, Caesarean delivery rates, and average length of stay suggest that the obstetrical care provided by the plan may not conform to best clinical practice.

In the plan, the non–birth-related obstetrical admissions averaged 35 percent of total births. This rate is much greater than the national normative admission rate of 13.8 percent. Delivery by Caesarean section accounted for roughly 20 percent of the births among plan enrollees, and this rate is less than the national average of 23.5 percent. In addition, the average length of stay following vaginal deliveries is 2.8 days in the plan, compared with the national norm of 2.2 days. At this time, unfortunately, the plan does not have quality or access data at the patient level to include in the matrix.

Clinical Enterprise Function

Contributors of the Clinical Enterprise Function

The unified model's engine is the clinical enterprise function. One of the major tasks of the clinical enterprise function is the selection of measures used in each cell of the performance matrix. It must also establish practice policies and target values for the performance measures. This function is responsible for development and implementation of utilization review protocols. Thus, the measures used in the performance matrix are entirely dependent on the clinical enterprise function.

As shown in Figure 15–14, clinical enterprise accepts inputs from information feedback in defining best clinical practice. To accomplish this goal, the clinical enterprise function must balance the results of empirical science with individual values. The information feedback and capacity engineering functions play necessary, but supporting, roles to the clinical enterprise function's defining of best clinical practice.

Like the information feedback function, the clinical enterprise function should be organized along product lines. This activity should focus on the health care of a population in a defined geographical area. The output from the clinical enterprise function comprises the following four products:

1. scope of benefits to determine health care services offered within the plan
2. clinical practice policy to guide providers in making clinical practice decisions

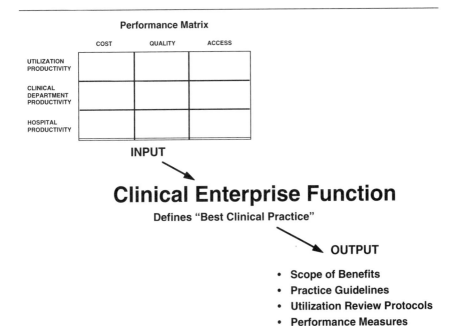

Figure 15–14 Inputs and Outputs of the Clinical Enterprise Function

3. utilization review protocols to manage clinical practice
4. measures to evaluate appropriateness of performance

Although these products are designed to affect different dimensions of the treatment process, they should be founded on the same scientific knowledge and should be generated from the same empirical approach.[12]

The most important contribution of the clinical enterprise function is development of an estimate of health care service volume to be consumed in a future time period. The service volume estimates are provided to the capacity engineering function.

Obstetrical Case Study: Clinical Enterprise

Accepting the information from the information feedback function, the plan staff made the following best clinical practice decisions:

1. Reducing non–birth-related obstetrical admissions from 35 percent of total births to 15 percent could save 1,037 admissions and 3,147 days of care per year.
2. Reducing Caesarean section deliveries from approximately 20 percent of births to 12 percent could reduce overall obstetrical average length of stay and save 1,236 days per year.
3. Reducing the average length of stay following vaginal deliveries from 2.8 to 2.35 days could save an additional 2,519 days.

In total, changing the quantity of services provided by reducing the non–birth-related admissions, the Caesarean delivery rate, and the average length of stay following vaginal birth has the potential for lowering the number of days of care provided by 6,902 days or 29 percent in the region.

Capacity Engineering Function

The usefulness of the quality/cost debate depends on valid measurement of the potential cost and quality ramifications of changes in clinical practice and the translation of potential savings into real savings. For example, most measurements of opportunity costs are invalid, because the analyses depend on average rather than marginal costs. If changing provider behavior would save 1,000 patient days at a cost of $1,000 per day, then $100,000 in cost savings could result—this statement is true only when all costs are variable. In reality, 80 percent or more of the $1,000 per day costs are determined by the fixed capacity, and these costs will not automatically change just because a patient no longer occupies the bed. Traditional analyses that rely on average costs do not provide decision makers with sufficient information to realize potential savings.

The potential cost savings from best clinical practice definitions can be realized only if capacity is sufficiently flexible. Defined as the facility, staff, and equipment necessary to provide a medical service, capacity determines the fixed or long-run costs for a plan or region. Therefore, once capacity decisions have been made, the cost of health care for the plan or region has largely been set. If the capacity decisions are incorrect or do not correctly anticipate changes in clinical volume, costs will be higher than necessary. To be competitive, a provider or plan must be capable of minimizing its capital investment in response to clinical volume changes.

As shown in Figure 15–15, capacity engineering accepts inputs from both the information feedback and clinical enterprise functions in determining the needed physical capacity of the health care delivery system. The goal of the capacity engineering function is to minimize the capital and related investments required to provide health care to a population. This function involves capacity-modeling tools that a manager can use to determine the needed size and staff of a facility and make technology investment decisions. Each tool is designed to permit interaction with clinical and administrative decision makers, and to reflect market differences. These sophisticated modeling techniques can be used to estimate the capacity responses associated with best clinical practice explorations. If clinical practice is changed, these tools guide the daily decisions required to achieve the resource savings.

In the past, the capacity engineering function has often been poorly done for two major reasons. First, capacity analyses have been based on historical volume; in this instance, sizing, staffing, and technology decisions were based on data that may not have reflected appropriate practice patterns. Second, the analytical techniques employed were unsophisticated. The sizing of facilities was based on average occupancy rates rather than on probability forecasts of patient demand. Inpatient staff were scheduled by average requirements per shift rather than hourly demand. Technology assessments and decisions were often focused on the hospital rather than the region. These practices resulted in excess capacity and increased costs.

Two major applications of operations research techniques to health care—facility sizing and staff scheduling—may be helpful in overcoming these problems.

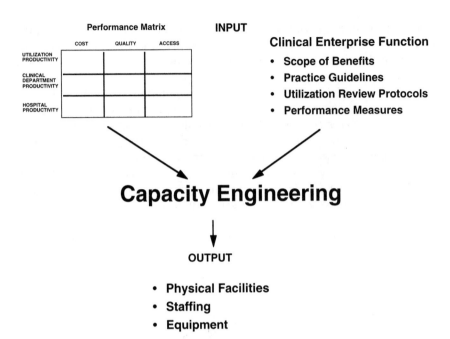

Figure 15–15 Inputs and Outputs of Clinical Enterprise Function

---

Facility Sizing

Patients admitted to a hospital fall into two broad categories: patients admitted for scheduled occurrences and patients admitted as a result of random events. For each category, different operations research tools apply. Most hospital admissions can be scheduled. The precise time a pregnant woman will enter labor is unknown, however, and birth-related obstetrical admissions are best modeled as random events. Patients admitted to the hospital through the emergency department are also considered random admissions. It is important to distinguish between these two types of admissions because different techniques are needed to manage them efficiently, and both techniques are needed to size the facility correctly.

Planning for Scheduled Patient Arrivals. The number of patients admitted to a hospital and the average daily census influence the number of beds necessary in a particular facility. Hospitals cannot build or operate for the average day, however. To control admissions, it is important to know the

Managing Quality under Cost Constraints 291

day-to-day *variations* in admissions and the average daily census. Knowing the variance in hospital occupancy allows the facility to determine the capacity that it needs to meet demand on the busiest days, while simultaneously achieving acceptable bed utilization. Tight control of admission scheduling makes it possible to maintain a careful balance between the need to meet demand and the need to maintain high bed utilization.

The purpose of admission scheduling is to control the admission process so that patients not requiring urgent care are scheduled away from the "peak" days to the "valley" days when the census is lower. Admission scheduling, therefore, is an attempt to reduce the variation in patient census and reduce the corresponding requirement to build and operate inpatient beds.

An admission scheduling system operates by having physicians classify their patients into categories such as emergent, urgent, scheduled, and call-in. A patient classified as urgent, for example, must be admitted within 3 days of the initial request. If not, the patient is reclassified to emergent and admitted immediately. Obviously, a well informed and cooperative physician staff is crucial to the success of an admission scheduling system.

A number of hospitals have successfully implemented admission scheduling systems. Typically, an admission scheduling system allows the hospital to operate at an average occupancy of 85 to 90 percent or more. Studies have documented significant resource and financial savings from these systems.[13]

Analytical Modeling for Random Patient Arrivals. The Poisson process is a mathematical model that has been shown to be particularly useful in planning for random patient demand, such as occurs in an obstetrical facility.[14] It was developed from probability and queuing theory. The use of this model requires data input on the arrival rate (the expected number of admissions or births per time period) and the service rate (average length of stay). Given these data, the Poisson process predicts bed requirements and the associated statistics on unit occupancy and patient turnaways. Application of the Poisson process to obstetrical facility sizing has shown that, as patient volume increases, the requirement for beds also increases; this increase does not follow a linear relationship, however. There is a significant economy of scale associated with high patient volume. Therefore, geographical areas that have multiple facilities can reduce their consumption of resources by consolidating services to a smaller number of facilities.

For example, if obstetrical services for 500 births per year are provided at four hospitals and the average length of stay is held constant at 2.5 days, each of the four hospitals must operate a total of 7 beds to meet demand 95 percent of the time. Therefore, a total of 28 beds must be operated. If, however, services were consolidated to one location with a volume of 2,000 births per year, a total of 20 beds would meet the same level of demand. This consolidation effort would save the cost of staffing and operating 8 beds every year. Typically, construction costs are in the range of $250,000 per bed, with operating costs in the range of $100,000 per bed per year or more. These 8 beds represent a reduction of approximately $2 million in construction costs and $800,000 savings per year in operating costs.

It is not unusual for the hospitals in large urban areas, like the obstetrical facilities, to provide duplicate services. It may be possible through provider networks to consolidate services within existing hospitals without reducing the medical services available to patients. This consolidation effort may increase utilization at a small number of locations and reduce the current maldistribution of medical resources. Consolidation also has the potential to improve service quality, because the management of a large number of similar cases at a single location leads to improved clinical judgment, more appropriate treatment, and more skillful technical execution of care.

Staff Scheduling

Operations research techniques have been successfully used to solve problems related to inpatient staffing and scheduling. These techniques focus on hospital requirements for registered nurses, licensed practical nurses, and medical technicians. Generally, the "staffing" problem is divided into three areas: staffing, scheduling, and allocation.[15] Staffing addresses the number of people to hire over a planning horizon of several months to 1 year. Scheduling determines who will work during each hour of each day; it focuses on a 2- to 4-week horizon as reflected in a posted work schedule. Finally, allocation focuses on near-term problems. For example, Monday morning at 8 A.M., the anticipated demand was X patients, but it is actually X plus 4. The workforce was expected to be Z workers, but it is actually Z minus 2 workers. How can today's workforce be matched to today's patient demand?

Conceptually, the goals of operations research techniques applied to the staffing-scheduling-allocation problem is similar to those of admission scheduling. Hospitals schedule their staff by shift to meet demand on the busiest days. The demand varies significantly by day of week and time of day. If the peak demand period requires 10 nurses and this period occurs on Thursday between 10 A.M. and 11 A.M., for example, the core staff for the day shift on Monday through Friday is established at 10 nurses. Fewer than 10 nurses may be required on Thursday before 10 A.M. and after 11 A.M. and on days other than Thursday, however. Operations research techniques can be applied to the staffing-scheduling-allocation problem in an attempt to match the staff on duty closely with patient demand by hour of the day and day of the week. Therefore, the traditional three-shift structure of days, evenings, and nights may be discarded in favor of a workforce with multiple shift lengths. For example, staff may have overlapping start and stop times with full-time staff working 8-, 10-, or 12-hour shifts and part-time staff working 4 hours.

Operations research applications to staffing-scheduling-allocation problems in hospitals have been successful in a number of institutions. Studies have documented significant personnel and financial savings from this implementation.[16] These applications can usually reduce staff requirements by 10 percent or more and save millions of dollars in yearly operating costs.

Obstetrical Case Study: Capacity Engineering

> Because obstetrical admissions are usually random events, admission scheduling has little effect on the obstetrical bed requirement. On the other hand, the region has considerable potential gains from the consolidation of obstetrical services. Obstetrical services are currently provided at six facilities, which greatly increases the "open the door" costs and requires more beds (and more staff). Consolidating the current facilities from six to two would reduce the requirement for obstetrical beds in the region from 96 to 84 (a 13 percent reduction in beds), even with no change in current practice patterns.
>
> If consolidation is not considered, significant savings are possible by changes in practice patterns. Reducing the rates of Caesarean delivery, admission for nondelivered obstetrical patients, and average length of stay would reduce the requirement for obstetrical beds in the region from 96 to 73 (a 24 percent reduction in beds).
>
> Combining the capacity engineering techniques with the clinical enterprise techniques described earlier reveals dramatic results. In this case study, the bed requirement decreased from 96 to 62 (a 35 percent reduction in beds). Table 15–1 displays in chart form both the independent bed requirement savings of changing practice patterns and of capacity engineering, as well as the potential savings of the interactive effect of doing both.

Table 15–1 Potential Savings from Applying Clinical Enterprise and Capacity Engineering

Region	Current Practice Patterns					Revised Practice Patterns			
	Total Admit	Act. ALOS	Beds Avail.	Beds Reqd.	Nurse Staff Savings	Total Admit	Exp. ALOS	Beds Reqd.	Nurse Staff Savings
HOSPITAL A	1,701	2.68	25	18		1,443	2.54	15	
HOSPITAL B	1,406	5.11	24	27		1,017	3.68	16	
HOSPITAL C	445	1.80	0	5		366	1.92	4	
HOSPITAL D	1,816	3.51	24	24		1,670	2.94	20	
HOSPITAL E	362	2.25	8	5		328	2.12	4	
HOSPITAL F	1,312	3.14	19	17		1,182	2.73	14	
TOTAL	**7,042**	**3.39**	**100**	**96**	—	**6,005**	**2.82**	**73**	**20%**
NCR HOSPITAL NORTH	3,667	3.92		50		3,052	3.06	34	
NCR HOSPITAL SOUTH	3,375	2.81		34		2,953	2.57	28	
TOTAL	**7,042**	**3.39**		**84**	**20%**	**6,005**	**2.82**	**62**	**30%**

Nurse Staffing Implications

Application of a staffing-scheduling-allocation tool is especially important in a random event process, such as obstetrical admissions, where fluctuations in workload make core scheduling by shift very inefficient. Although the randomly arriving workload can not be accommodated through admission scheduling, it can be anticipated. Therefore, efficiencies are possible if staff can be scheduled to meet the anticipated demand.

An estimate of the potential savings in nurse staffing as a result of the changes in practice patterns and capacity engineering is shown in Table 15–1. Changes in nurse staffing requirements are more difficult to estimate than changes in bed requirements because practice pattern changes will most likely increase the nurse to bed staffing ratio; as a result, the decrease in nursing requirements will be smaller than the decrease in beds. Preliminary estimates suggest that a 20 percent reduction in nurse staffing is possible if either practice pattern or capacity engineering changes are implemented separately, and a 30 percent reduction is possible if they are used together.

CONCLUSION

The capability to manage the cost/quality relationship that constrained resources demand requires a structured approach to determining and implementing best clinical practice. The integration of information feedback, clinical practice decisions, and capacity management in a unified model can help to define and manage the clinical and administrative decisions that determine both the cost and quality of care. With increasing demands for both clinical and financial accountability, the unified model provides a structure for improved dialogue. Only with such an improved dialogue can the interested parties come to grips with the subjective nature of the cost/quality relationship.

NOTES

1. D. Eddy, Applying Cost Effectiveness: The Inside Story, *JAMA* 268, no. 18 (1992):2575–2582.
2. D. Eddy, Broadening the Responsibility of Practitioners, *JAMA* 269, no. 14 (1993):1849–1855.
3. A. Donabedian, *Explorations in Quality Assessment and Monitoring: The Definition of Quality and Approaches to Its Assessment*. Vol. I (Ann Arbor, Mich.: Health Administration Press, 1980), 4–5.

4. Ibid., 5.
5. The term *clinical enterprise* is derived from conversations with David Eddy concerning his Enterprise Model from February 93 to April 93 in Washington, D.C.
6. R.I. Lee and L.W. Jones, *The Fundamentals of Good Medical Care* (Chicago, Ill.: University of Chicago, 1993).
7. J.E. Wennberg, Dealing with Medical Practice Variations: A Proposal for Action, *Health Affairs* 3 (1984):6–32.
8. Donabedian, *Explorations in Quality Assessment and Monitoring.*
9. R. Penchansky and J.W. Thomas, The Concept of Access: Definition and Relationship of Consumer Satisfaction, *Medical Care* 19, no. 2 (1981):127–140.
10. Donabedian, *Explorations in Quality Assessment and Monitoring.*
11. See Donabedian, *Explorations in Quality Assessment and Monitoring,* for a more complete discussion of the relationship among structure, process, and outcome measures.
12. D.M. Eddy, *A Manual for Assessing Health Practices and Designing Practice Policies: The Explicit Approach* (Philadelphia: American College of Physicians, 1992).
13. W.M. Hancock et al., Admission Scheduling and Control Systems, in *Cost Control in Hospitals,* ed. J.R. Griffith et al. (Ann Arbor, Mich.: Health Administration Press, 1976), 150–187.
14. P. Cowan and K. Roth, Determining Maternity Case Load by Means of a Poisson Process, *British Journal of Preventive Medicine* 18 (1964):105–108.
15. D. Warner et al., *Decision Making and Control for Health Administration* (Ann Arbor, Mich.: Health Administration Press, 1984).
16. A.S. Shorr, *The Optimizer: A Policy and Productivity Evaluation Methodology for Acute Care Hospitals* (Woodland Hills, Calif.: Arthur S. Shorr & Associates, 1988).

» Chapter 16 «

Beyond the Hospital Door

Vahé A. Kazandjian and Elizabeth L. Sternberg

KNOWLEDGE is power.

Today, health care reformers—professionals, public advocates, politicians, and others—increasingly recognize the importance of good data. Appropriate and timely information is the key to sound public policies on an issue both private and public, both scientific and emotional, both quality-conscious and cost-conscious. Health care policy makers are already promoting the use of carefully designed health care databases to implement quality assurance and improvement programs, to develop effective cost containment measures, to explore medical practice style differences, and to study medical outcomes.

Information gathering and reform initiatives are currently traveling two paths: one that seeks national action on health care and one that favors the development of a new information generation at the statewide level. What shape will the organization of health care services take in the coming years? Will there be a shift toward a systematic quantification of the appropriateness of care? Will the public mandate for accountability translate into reliable, timely, and useful information? Or will the public be subjected to sensational exposés of confusing and narrowly viewed moments in health care practice?

COMMON METHODS OF INQUIRY

The methods used to analyze health and health services utilization data (primarily hospital discharges) in various statewide databases have been quite similar. One still popular approach to data analysis is the use of Small area variation analysis (SAVA) to identify variations in the rates of hospitalizations and receipt of surgical care among defined geographical or market areas, usually within a state. The SAVA technique is primarily epidemiological in nature, since it involves calculating areal rates, adjusting the denominators for the appropriate group at risk, comparing rates across areas and over time, and suggesting possible consequences of the interarea variation in rates to quality of care.

Although introduced in the 1970s, the observation of variation in the use of health care services or the rates of surgical care has historical antecedents. In 1856, Dr. William A. Guy, a physician at Kings College Hospital in London, reported that inhabitants of the Parish of St. Mary-le-Strand were hospitalized 325 times more frequently than were those of the district of Marylebone. His explanation? He characterized St. Mary-le-Strand residents as

> an increasing class of working men, in receipt of good wages, who are in the habit of applying to hospitals as a matter of course, even for trifling attacks of illness, to say nothing of those which sometimes follow immediately on expensive acts of self indulgence.[1]

It is tempting to say, a century later, that similar patterns persist, perhaps at a higher societal cost.

Also in England, the variation in rates of surgical care was recognized and documented more than 50 years ago. Glover identified 10-fold variation in the rates of tonsillectomy among rural communities.[2] Although variation analysis subsequently became dormant for many decades, the technique was resurrected during the 1950s. In 1959, Lembcke effectively demonstrated that epidemiological techniques of data analyses could be used as part of an education program for medical staff to reduce rates of surgical procedures.[3] By integrating the findings of medical audits into a medical education program, Lembcke reduced rates of appendectomy and tonsillectomy by half. Since then, Wennberg has popularized the SAVA method and, in the past two decades, has demonstrated that this population-based, epidemiological technique can assist policy makers in the search for timely and informative data.[4]

The SAVA technique is not without its limitations, however. Because it relies on previously collected data and such data are predominantly inpatient care–specific, analysis is often limited to inpatient hospital procedures. As more and more types of care are shifted to the outpatient setting, caution must be taken not to attribute differences between providers to practice patterns when the differences actually represent greater use of the outpatient setting. Ultimately, the full potential for SAVA will be achieved through the development of databases that integrate hospital, outpatient, and ambulatory care services.

STATE LEVEL INITIATIVES

A variety of different state level database systems have been created and used in the documentation of the local population's health and health services utilization profiles. The primary goal of these initiatives is to enable the implementation of cost-effective strategies to improve quality care, reduce unnecessary care, and contain cost.

Prior to the groundswell of state health care reform activity in the late 1980s, the Maine Medical Assessment Foundation (MMAF) was established as a privately sponsored database program with partial support from the Robert Wood Johnson Foundation. By 1990, with additional support, the MMAF had established eight medical specialty study groups to explore regional variations in hospital-based medical care.[5] The study groups employed the SAVA technique to identify statistically significant excessively high or low rates of certain procedures, as well as rates of hospitalizations across geographical regions. Among the focused analyses were rates of hysterectomy, prostatectomy, and lumbar disc surgery; paediatric hospital admissions; adult pneumonia hospitalizations; and rates of cataract surgery. Data analysis revealed that variations persisted even after adjustment for differences in demographic characteristics, insurance status, and income. Because the study groups consisted of specialists from across the state, their findings and recommendations influenced practice patterns among many providers throughout Maine. The feedback of the results to physicians often culminated in a change in rates. Although the rates decreased in most instances, there was no convincing evidence that the quality of care had improved.

Precise data can be an extremely powerful tool for reducing the incidence of unnecessary surgical procedures or hospital admissions, even in the absence of agreed upon medical practice guidelines or standards of care. Appropriate changes in practice styles translate into substantial economic savings. Estimated 1989 savings due to the MMAF data analysis and information dissemination equaled $6 million, while total savings for the period 1985 to 1986 were estimated to be $21 million. These reflect only savings in hospital charges and exclude savings resulting from the avoidance of physician fees and the elimination of a wide range of indirect costs (e.g., patient harm, impairment of quality of life) that would substantially increase the true savings generated by the project.

Other states have also recognized the need for developing nonhospital databases. The Wisconsin Ambulatory Medical Care Survey is modeled after the National Ambulatory Medical Care Survey (NAMCS). The Wisconsin survey is designed to produce a comprehensive ambulatory care utilization database. Three surveys have been conducted thus far (1983–

1984, 1986–1987, and 1989–1990). Information compiled through the survey includes profiles of patients, physician practice style description, patient visit rates, and profiles of most frequent medical diagnoses.

The Iowa Health Policy Corporation (HPC) was established to study the frequency and type of hospital outpatient procedures that involve health insurance claims. Data are aggregated according to procedures using diagnosis-related group (DRG) codes, as well as by characteristics from individual patient records.

In the only all payer state in the nation, the Maryland General Assembly enacted sweeping health care reform legislation in 1993. Among other mandates, the legislation established a new statewide organization to study not only variations in hospital utilization rates, but also certain aspects of quality of care. The Health Care Access and Cost Commission, which will have a wide range of responsibilities, must establish an integrated hospital and ambulatory health care database, using claims organized through standardized claim forms and electronic processes. The commission is charged with analyzing all data to report rates of utilization, changes in services, and variations in provider fees. These findings will be published annually and will serve as the basis for the commission's recommendations on cost containment and quality improvement strategies for the state.

Other state initiatives include projects for the creation of databases for hospital outpatient procedures (Illinois), office-based medical services (New Hampshire), and physician charges for ambulatory services (North Dakota).

NATIONAL INITIATIVES

It is practically impossible not to stumble on an analysis of health care reform daily, in the printed or electronic media. Reform is the agenda, and the need for change seems widely accepted. Three fundamental aspects of health care delivery constitute the framework for change: (1) access to care, (2) cost of care, and (3) quality of care. The relationship among these three components of a conceptual framework has been proposed and analyzed in the classic works of Donabedian.[6] In the 1990s, however, this conceptual organization of the health care delivery system has reached the policy level, nationally and on state levels. Thus, the relationship between research (academic and applied) and policy is now an intimate one. It can be hoped that, as a result, policy decisions will increasingly be based on more objective, scientifically supported, quantifiable observations and findings. To improve the relevance of policy decisions, a better understanding of the scientific methods of inquiry will be necessary for policy makers. Therefore, a familiarity with epidemiological methods and concepts of performance analysis is of fundamental relevance.

Any proposed reform model is likely to give accountability a pivotal role. In the 1990s the most popular form of accountability regarding the performance of a health care organization is that of a "report card" or "indicator system." Until the content and the format of presentation are refined, it is possible that both the general public and the purchasers of care will be confused and will vary in their interpretation of the report card's message. When the distilled, comprehensible, and reliable information about the appropriateness or efficiency of services at various health care organizations is achieved, it is expected that the reader of the report card will make more reasonable and educated decisions. Until then, organizations that have an apparently less desirable performance profile may unjustly suffer from the public's scrutiny. Indicator systems may vary in their targeted audiences. For example, if the target audience is the health care professional eager to know more about his performance and anxious to ameliorate it, the indicator rate feedback to the professional may serve a constructive purpose.

REORGANIZING THROUGH HEALTH CARE NETWORKS

That health care resources are limited is slowly sinking into the general consciousness. When it comes to medical interventions, the popular request has always been "more, and abundantly more." As health care costs spiral upward and it becomes clear that resources may be inefficiently used, a distinct shift in thinking is taking place. How can society provide access to health care, maintain or even enhance its quality, and regulate the escalation of its cost?

One proposed approach is to create a system of health care organizations ranging from inpatient secondary and tertiary care centers to outpatient service organizations, to home care agencies, to primary care clinics. The idea is not a new one. In its most fundamental form, the proposed network seems very similar to the 1950s concept of regionalization of services. In short, each element of the system would fulfill a specific task (thus eliminating redundancy); the interaction between the different elements (inpatient, ambulatory, home care, long-term care, rehabilitation services) would be organized, supervised, and in some form purchased by a separate entity. The latter component of the proposed network plan is distinctly new. The primary function of the purchasing entity seems to be to control cost variation through a more uniform cost allocation to types of services. The secondary function may be the "leveling of the playing field" with regard to access to care.

The potential success of the network scheme is based upon a number of core assumptions. First, it must be assumed that there is or will be a reliable, comprehensive, and ongoing assessment of population need. In fact, the argument about networks or regionalization of services holds only if the planning for that system is based on population need for different types of services. There must be reliable information on population health status, incidence rates of acute illnesses, prevalence rates of chronic conditions, and valid projections for epidemics. Thus, the establishment of epidemiological intelligence will be vital to any attempt of regionalization.

Second, barriers to access must be diminished, especially for preventive and public health services. In fact, it may be argued that the cost:benefit analysis is overwhelmingly favorable when cost of screening and prevention are compared with the cost of treatment. The dedicated commitment from leaders and society to minimize barriers to access through universal basic coverage, outreach programs, and the establishment of low-cost high-impact primary care services within communities will alleviate this problem.

Third, through the collaborative effort of all parties—patients, insurance carriers, small business representatives, major purchasers of care, the government, and the medical profession—a consensus will be reached about the uniform strategy of action.

Finally, there will be a reliable, valid, and accepted set of strategies to ascertain, assess, evaluate, and monitor the goodness of the services. These strategies must be built upon the principles of epidemiological population-based techniques complementing the guidelines from management engineering theories, where population (customer) need is met with the most efficient approach.

Comprehensive change, however, may occur even without policy mandates. The "Pandora's box" of health care information is open. Health care consumers may eventually demand to know more about the quality and cost of care than they ever have in the past.

CLOSING THOUGHTS

In the 1990s, it seems that the route to health care improvement is paved with good intentions. Not only does the serpentine path of the route to improvement seem clearer on the map, but the tools for extending its paving are increasingly available to the interested. From the rediscovery of epidemiological techniques to the recognition of the public's right to know, value, and judge health and disease care, the spectrum has been broadened. A most fundamental coverage of the broader spectrum may be the pursuit of a continuum of care model, where population health status is evaluated in relationship to the performance of the curative, preventative, and palliative attributes of our health care system.

In the hands of globally educated, more caring, and better guided providers, the relationship of sickness and health in the United States will undoubtedly change in the twenty-first century.

NOTES

1. B.A. Barnes, Population-Based Unit Analysis of Health Care, in *Regional Variation in Hospital Use*, ed. Rothberg (Lexington, Mass.: Lexington Books, 1982), 188.
2. J.A. Glover, The Incidence of Tonsillectomy in School Children, *Proceedings of the Royal Society of Medicine* (1938):1219–1236.
3. P.A. Lembcke, Measuring the Quality of Medical Care through Vital Statistics Based on Hospital Service Areas. 1. Comparative Study of Appendectomy Rates, *American Journal of Public Health* (1952):276–286.
4. J.E. Wennberg and A. Gittelsohn, Small Area Variation in Health Care Delivery, *Science* 182 (1973):1102–1108.
5. R.B. Keller, Maine Program Analysis Small Area Variations, *The AAOS Bulletin*, July (1987):9–12.
6. A. Donabedian, *Aspects of Medical Care Programs* (Cambridge, Mass.: Harvard University Press, 1976).

Index

A

Access, definition of, 279–280
Access measures, health plan performance measures, 92–93
Accountability. *See* Public accountability
Administrative data sets
 in health plan performance measures, 87, 93–94
 types of, 93–94
Agency for Health Care Policy and Research, 90
AIDS, relationship to tuberculosis, 231, 233
Alpha, 45
Ambulatory care
 administrative aspects of, 180
 ambulatory encounters, lack of understanding of, 180
 ambulatory encounter system (AES), 180–181
 community-oriented perspective, 178–179
 forces related to, 174–175
 and increased admissions, 177
 space allocation for, 176
Ambulatory care quality
 claims-based data studies, 185–189
 classification tools for measurement of, 179–181
 clinical indicators, 189
 measures for improvement of process, 181–182
 patient satisfaction surveys, 182–184
 population-based approach, 173–174, 177
 protocols, 189–191
American Medical Review and Research Center, 90
APACHE Medical Systems, 145, 150, 154
Appropriateness
 and best clinical practice, 271–273
 and cost, 271–272
 definition of, 7–8, 270–271
 and quality, 272–273
Attributable risk, 17
Austin Project, 179

B

Best clinical practice, 271–273
 and appropriateness, 271–272
 definition of, 271
 relativity dilemma of, 273

unified model for health care
    delivery, 273–295
Beta, 45
British health care system
    clinical audit in, 203–207
    clinical guidelines, 205–206
    continuing education of physicians,
        207–208
    duties of physicians, 210
    early years of, 197–199
    forces influencing quality, 200–201
    general management, beginning of,
        199–200
    medical education,
        recommendations for, 208–209
    multidimensional performance
        assessment, 212–216
    1990 reform, 200
    poor practice, handling of, 210–211
    practitioner performance,
        assessment of, 218–219
    purchasing for quality, 201–203
    recommendations for future,
        208–209
    service performance, assessment of,
        211–218

C

Calendars, assessment for quality, 68
Cardiac Surgery Reporting System,
    241–251
    areas studies, 242–243
    audience for data, 244
    data collection, 242
    hospital evaluation, 245
    patient information, 247–248
    physician-specific performance
        data, 246
    problems related to, 250–251
    purposes of, 241
    results of, 243
    usefulness of, 248–250
Case-control methods, 23

Case-finding methods, 57
Center for Health System Studies, 97
Claims-based data studies
    ambulatory care quality measure,
        185–189
    categories of care tracked by, 186,
        188
    nature of, 181
    small area variation analysis
        studies, 185
Classification tools
    ambulatory care quality measure,
        179–181
    coding systems for disease, 180–181
Cleveland Health Quality Choice
    Coalition, 144–169
    attributes of program, 147
    Communications Committee, 147
    database, use of, 166–168
    intensive care unit data, 150–154
    medical-surgical outcome data,
        154–158
    members of, 145–146
    misinterpretations of data, reasons
        for, 160–161
    objectives of, 146
    patient satisfaction data, 148–150
    Quality Information Management
        Corporation, 147, 148–149
    report format task force, activities
        of, 158–163
    training workshop, 159–160
Clinical indicators
    ambulatory care quality measure,
        189
    nature of, 182
Clinical pathways
    elements in system, 118
    format approach to, 114–117
    performance indicators, 119
    team, use of, 111–113
    use in design of care, 110–119
Clinical practice
    best clinical practice, 271–273

practice guidelines, 138
Clinton administration, health care reform, 84
Codman, Dr. Ernest, 56
Community care networks, 254–266
  areas of accountability, 256–259
  assessment of community health needs, 259–260
  assessment of health care system, 260
  challenges related to, 264–266
  community, definition of, 264–265
  Community Health Intervention Partnerships (CHIPs), 262–263
  community-oriented primary care, 262
  Communitywide Health Improvement Learning Cooperative, 264
  and customer-driven quality, 257–258
  data analysis in, 258
  data sources for, 265–266
  evaluation in, 261
  goals of, 256
  Hospital Association of Pennsylvania model, 263
  Hospital Community Benefits Standards Projects, 263–264
  interventions, selection of, 261
  leadership of, 257
  objective setting, 261
  partnership development, 260
  Planned Approach to Community Health (PATCH), 264
  prevention efforts, 258
  priorities, establishment of, 260–261
  system thinking in, 258
  teamwork in, 257
  values, importance in, 266
Community Health Intervention Partnerships (CHIPs), 262–263
Community oriented health status management, 185

Communitywide Health Improvement Learning Cooperative, 264
Condition-specific measures, health plan performance measures, 88, 90–91
Consortium Research on Indicators of System Performance, 97–98
Content analysis, of quality reporting documents, 70–71
Continuous quality improvement
  programs, 57–58
  importance of, 172–173
  and Japanese, 124
Cost
  and appropriateness, 271–272
  cost constraint, 272
  cost/quality trade-off, 272
  meaning of, 269–270
Cost containment. *See* Unified model of health care delivery
Cost-effectiveness, and quality, 7
Critical paths
  factors for success of, 108
  nature of, 182
Cross-sectional survey, 40
Culture of organization
  assessment for quality management, 62–68
  questionnaire assessment of, 65–67
Current Procedures and Terminology, 180
Customer satisfaction analysis, survey research, 47–50

D

Darling v. Charleston Community Memorial Hospital, 57
Dartmouth Co-op Charts, 183–184
Database
  for Cleveland Health Quality Choice Coalition, 166–168
  state level systems, 299–300

Data collection
  special instruments for, 96
  in survey research, 43–45
Data sources
  for community care networks, 265–266
  for epidemiology investigation methods, 19–21
  for health plan performance measures, 87–88, 93–96
Data thresholds, use of, 132, 134–137
Data trending, use of, 137–138
Develop and Evaluate Methods to Promote Ambulatory Quality project, 186, 187–188, 189

E

Employee Health Value Survey, 91
Enrollee surveys, in health plan performance measures, 87, 95–96
Epidemics, dimensions of, 14
Epidemiological model, in unified model of health care delivery, 276–279
Epidemiology, meaning of, 14
Epidemiology investigation methods, 21–23
  case-control methods, 23
  data sources in, 19–21
  longitudinal methods, 21–23
  problem definition in, 19–20
  system characteristics in, 20–21
Epidemiology measures, 13–21
  incidence rate, 16
  issues related to rate assessment, 15–16
  prevalence rate, 16–17
  risk measures, 17
  sensitivity and specificity of test, 17–18
  validity and reliability of, 15

Evaluation of health care
  epidemiology investigation methods, 21–23
  epidemiology measures, 13–21
Evaluation matrix, in unified model of health care delivery, 280–282
Executives
  chief quality officer, 76–77
  quality meetings of, 80–81
Experimental survey, 41
  quasi-experimental survey, 41

G

Greater Cleveland Hospital Association, 144–145
Group Health Association of America, health plan measurement system, 91

H

Health Care Access and Cost Commission, 300
Health Care Financing Administration, 29, 94, 99, 189
Health care reform
  Clinton administration, 84
  forces related to reform, 239–240
  reorganization through health care networks, 301–302
Health Plan Employer Data and Information Set (HEDIS), 98–99, 191, 240
Health plan performance measures, 86–103
  access/use of service measures, 92–93
  condition-specific measures, 88, 90–91

Consortium Research on Indicators of System Performance, 97–98
data capabilities, 102
data sources in, 87–88, 93–96
and external auditing, 102–103
Health Plan Employer Data and Information Set (HEDIS), 98–99
health status measures, 91–92
high-priority topics, 89
InterStudy outcomes management system, 96–97
limited set of, 101
Massachusetts Purchaser/HMO Collaborative effort, 98
Medicare performance measurement demonstration, 99
public reporting of, 100–103
refinement over time, 88
requirements of measurement system, 87
risk adjustment for, 102
satisfaction measures, 91
Health plans
characteristics of, 85–86
in-network plans, 86
types of, 85
Health status measures, health plan performance measures, 91–92
High-risk activities, types of, 130
Hospital Association of Pennsylvania model, 263
Hospital Community Benefit Standard Project, 257
Hospital Data Quality Overview Report, 165
Hospital Quality Trends Patient Judgment System, 148
Hospital Standardization Program, 56, 123
HSQ-12, 92
Human resources management, and survey research, 51

Hypotheses
null, 45
in survey research, 39–40

I

Improvement teams, 79–80
Incentives, and quality, 5
Incidence rate, 16
Indicators
clinical indicators, 182, 189
common uses of, 26
construction of, 28–32
defining indicator, 27–28
indicator statements, 132
indicator worksheet, 133–134
outcome indicators, 126, 282
performance indicators, 126–134
process indicators, 126–127, 282
quality indicators, 26
sentinel event indicators, 131
testing of, 27–28
usefulness of, 26–27
See also Performance indicators
Information
assessment of, 73–74
levels in health care settings, 72
In-person surveys, 44–45
Institute for Healthcare Improvement, 257
Intensive care unit data, 150–154
International Classification of Disease, 155, 180–181
InterStudy outcomes management system, 90, 96–97
Iowa Health Policy Corporation, 300

J

Joint Commission on Accreditation of Healthcare Organizations, 123–124
formation of, 56
Indicator Measurement System, 240

L

Longitudinal methods, 21–23
  longitudinal survey, 40–41
  nature of, 21–22

M

Mail surveys, 45
Maine Medical Assessment Foundation, 299
Managed Care Association, 97
Market research
  concerns related to, 50–51
  elements of, 50
  surveys, 50–51
Massachusetts Purchaser/HMO Collaborative effort, 98
Media, as public information source, 249
Medical records, in health plan performance measures, 87, 95
Medical-surgical outcome data, 154–156
Medicare, mortality reports, 29–30
Meetings
  of improvement teams, 79–80
  of quality executives, 80–81

N

National Ambulatory Medical Care Survey, 299–300
National Committee for Quality Assurance, 99, 191
  national report card, 101, 102–103, 240
National Demonstration Project on Quality Improvement in Health Care, 124
National Health Service. *See* British health care system.

Newsday, Inc. vs. The New York State Department of Health, 240
Null hypothesis, 45
Nurse staffing, in unified model of health care delivery, 295

O

Organization
  information for quality reporting, levels for, 77–79
  organizational design, definition of, 55
  planning for future, 75–76
  reporting relationships in, 76–77
  senior executive quality officer, 76
Outcome indicators
  nature of, 126, 282
  vs. process indicators, 126–127

P

Paradox, resolution of, 76
Patient outcomes reporting. *See* Cleveland Health Quality Choice Coalition
Patient satisfaction surveys
  ambulatory care quality measure, 182–184
  customer satisfaction feedback, 63
  issues addressed in, 182
  nature of, 181
  patient satisfaction data, 148–150
  time for completion of, 183
Patients, views of quality, 125
Peer Review Organizations, 57
Pennsylvania Health Care Cost Containment Council, 240
Performance Assessment Committee, 99
Performance benchmarks, 49

Performance indicators
  communication of information
    about, 140, 142
  data evaluation for, 132–139
  development of, 127–129
  identification of meaningful
    indicators, 127–129
  outcome indicators, 126
  process indicators, 126–127
  rate-based indicators, 132
  rate construction, 32–36
  reliability of, 30–32, 130–131
  selection of useful indicators,
    129–130
  validity of, 28–30, 130–131
Pilot study, 46–47
  issues addressed in, 47
Plan-do-check-act (PDCA) cycle, 185,
  194–196
  act phase, decisions made in,
    195–196
  check in, 195
  methodologies in, 194–195
  plan in, 194
Planned Approach to Community
  Health (PATCH), 264
Population, meaning of, 32
Practice guidelines, purpose of, 138
Prevalence rate, 16–17
Primary data, 50–51
Probability sample, 46
Problem definition, in
  epidemiological investigation,
  19–20
Problem solving
  assessment of organizational
    problem solving, 70–72
  in quality management, 68–72
  styles of, 69
Process indicators
  nature of, 126–127, 282
  vs. outcome indicators, 126–127
Productivity, in unified model of
  health care delivery, 275–276

Product lines, development of,
  282–285
Professional Standards Review
  Organization, 56–57
Protocols
  ambulatory care quality measure,
    189–191
  characteristics of, 189–190
  criteria for content of, 190–191
  implementation difficulties,
    190–191
  nature of, 183
Public accountability
  of American Hospital Association,
    254–255
  community care networks, 254–266
  meaning of, 253
  and report card method, 253–254
Public health policy
  medical paradigm related to, 223
  program decisions, basis of,
    223–224
  and sexually transmitted disease,
    222–223
  See also Tuberculosis
Public reporting, of health plan
  performance measures, 100–103

Q

Quality
  components of, 272–273
  and consumer satisfaction, 6–7
  and cost-effectiveness, 7
  health care practitioner's views of,
    125
  and incentives, 5
  in industry, 5–7
  patient views of, 125
  purchaser's views of, 126
  quantification of, 5–6
  strategic approaches to, 108–109
  and use of indicators, 26

Quality health care
  definitional problems, 3–4
  and methods of care, 9
  and value of care, 10–11
Quality improvement
  levels of involvement in, 172–173
  and quality management, 61–62
Quality indicator, meaning of, 26
Quality Information Management
    Corporation, 147, 148–149
Quality management
  culture assessment in, 62–68
  definition of, 62
  historical background, 55–59,
    123–124
  information/measurement
    assessment, 72–75
  organizational assessment in, 61
  phases in, 59–60
  problem solving in, 68–72
Quality planning, 81
Quality reporting
  governance level, 81
  nature of, 80
Quality reporting documents
  content analysis of, 70–71
  forms for, 71
  information flow in, 77, 79
Quasi-experimental survey, 41
Questionnaires
  construction of items, 65–66
  design of, 66
  distribution of, 66
  nonresponse patterns, 66–67

R

RAND HMO Consortium, health plan
    performance measures, 88, 90
RAND Medical Outcomes Study, 92
Random error, for rate, 35–36
Rate-based indicators, 132
  types of, 136

Rates
  elements of, 32–33
  errors related to, 33–36
    random error, 35–36
    systematic error, 34–35
  errors related to understanding of,
    32
  formula for, 32–33
  incidence rate, 16
  indicator acceptability, 36
  issues related to rate assessment,
    15–16
  prevalence rate, 16–17
Receiver operator curve, 165
Relative risk, 17
Reliability of test
  determination of, 19
  meaning of, 15
  and performance indicators, 30–32
  of performance indicators, 130–131
  survey research, 42–43
  test-retest reliability, 43
Report cards
  accountability of, 253–254
  Cardiac Surgery Reporting System,
    241–251
  nature of, 96
  questions related to, 240
  types of performance systems, 240
Reporting relationships, 76–77
Risk-adjustment methodology,
    249–250
Risk measures
  attributable risk, 17
  relative risk, 17

S

Sample
  and power of statistical test, 46
  probability sample, 46
  sample size estimation, 45–46

sampling plan, 46
Satisfaction measures, health plan performance measures, 91
Satisfaction surveys
   pragmatic uses of, 64
   problems of, 63, 64
Secondary data, 50–51
Senior executive quality officer, 76
Sensitivity of test, 17–18
Sentinel event indicators, 131
Small area variation analysis studies, 185, 297–298
   limitations of, 298
   nature of, 297–298
Specificity of test, 17–18
Staff scheduling. *See* Unified model of health care delivery
Survey research, 38–52
   cross-sectional survey, 40
   customer satisfaction analysis, 47–50
   experimental survey, 41
   and human resources management, 51
   in-person surveys, 44–45
   instrument credentialing for, 42–43
   longitudinal survey, 40–41
   mail surveys, 45
   market research, 50–51
   pilot study, 46–47
   purposes of, 38–39, 52
   quasi-experimental survey, 41
   reliability and validity in, 42–43
   research hypotheses in, 39–40
   sample size estimation, 45–46
   sampling plan, 46
   in technological assessment, 51–52
   telephone surveys, 44
   variables, measurement of, 42
Systematic error, for rate, 34–35
System characteristics, in epidemiology investigation methods, 20–21

T

Teams
   in clinical pathway approach, 111–113
   in community care networks, 257
   improvement teams, 79–80
   nature of team approach, 109–110
   roles in, 110
Technological assessment, survey research in, 51–52
Telephone surveys, 44
Test-retest reliability, 43
Total quality management (TQM), 52
Tuberculosis
   at-risk groups, 224, 226, 233
   contagious aspects of, 225
   Directly Observed Therapy, 234
   etiology of, 225
   incidence in United States, 224, 228
   public program, evaluation of, 234–235
   registries for, 225
   relationship to AIDS, 231, 233
   symptoms of, 225
   tuberculosis wave, 226–228, 229–230
Type I error, 45
Type II error, 45
TyPEs (Technology of Patient Experience), 97

U

Unified model of health care delivery, 273–295
   capacity engineering, case example, 293
   capacity engineering component, 274, 288–295
   clinical enterprise, case example, 288
   clinical enterprise component, 274, 286–288

cost/quality/access parameters, 279–280
epidemiological framework, 276–279
evaluation matrix in, 280–282
information feedback, case example, 285–286
information feedback component, 274, 275–286
nurse staffing, 295
productivity in, 275–276
product lines, establishment of, 282–285
random patient arrivals, analytical modeling for, 291–292
scheduling of patients, 290–291
size of facility, 290
staff scheduling, 292–293
United HealthCare Corporation report card, 100
Use of service measures, health plan performance measures, 92–93

Utilization and Quality Control Peer Review Organizations, 57

V

Validity of test
  meaning of, 15
  and performance indicators, 28–30
  of performance indicators, 130–131
  sensitivity and specificity measurement of, 17–18
  survey research, 42–43
Value of care, and quality health care, 10–11

W

Wisconsin Ambulatory Medical Care Survey, 299